W9-AFE-810

Ecological Futures

TRILOGY ON WORLD ECOLOGICAL DEGRADATION
by
Sing C. Chew

World Ecological Degradation: Accumulation, Urbanization, and Deforestation, 3000 B.C.–A.D. 2000

The Recurring Dark Ages: Ecological Stress, Climate Changes, and System Transformation

Ecological Futures: What History Can Teach Us

Ecological Futures

What History Can Teach Us

Sing C. Chew

A Division of
ROWMAN & LITTLEFIELD PUBLISHERS, INC.
Lanham • New York • Toronto • Plymouth, UK

AltaMira Press
A division of Rowman & Littlefield Publishers, Inc.
A wholly owned subsidiary of The Rowman & Littlefield Publishing Group, Inc.
4501 Forbes Boulevard, Suite 200
Lanham, MD 20706
www.altamirapress.com

Estover Road
Plymouth PL6 7PY
United Kingdom

British Library Cataloguing in Publication Information Available

Library of Congress Cataloguing-in-Publication Data

Chew, Sing C.
Ecological futures : what history can teach us / Sing C. Chew.
p. cm.
ISBN-13: 978-0-7591-0453-2 (cloth : alk. paper)
ISBN-10: 0-7591-0453-0 (cloth : alk. paper)
ISBN-13: 978-0-7591-0454-9 (pbk. : alk. paper)
ISBN-10: 0-7591-0454-9 (pbk.)
eISBN-13: 978-0-7591-1223-0
eISBN-10: 0-7591-1223-1

1. Environmental degradation—History. 2. Global environmental change—History. 3. Ecological disturbances—History. 4. Climatic changes—History. 5. Sustainable development—History. I. Title.

GE149.C538 2008
363.7009—dc22 2008001993

Printed in the United States of America

∞™ The paper used in this publication meets the minimum requirements of American National Standard for Information Sciences—Permanence of Paper for Printed Library Materials, ANSI/NISO Z39.48-1992.

For three good friends
Beau (1993–2007), Matthias Gross, and Gunder Frank (1929–2005)

Contents

Figures and Tables

FIGURES

TABLES

Preface

This final volume brings to an end my long-term project to understand the dynamics of the relationship between Culture and Nature over world history. Started nearly a decade ago to study global deforestation, my examination has widened over the years. After so many years of trying to understand this historical relationship, I have come to the conclusion that history has much to teach us. Unfortunately, we do not rely on history nearly enough. I believe we should, and to this end I find the phrase "history as the teacher of life" to be a useful guide. That is the message I want to convey to the reader, and it continues the conclusion made in my first volume, *World Ecological Degradation,* that unless we learn from our common history, we as a human community because we do not learn from our common historical past will repeat the same mistakes that other prior civilizations have made.

This third volume follows the argument made in volume 2, *The Recurring Dark Ages,* that social systems do slip into crisis once natural system limits are breached, and over world history we have experienced at least three of these system crises. In this third volume, I suggest that perhaps we are already entering another crisis period if we counterpose our present conditions and trends with the historical patterns of past Dark Ages. In *The Recurring Dark Ages,* I left the reader with an ambivalent conclusion about our common future. Various audiences in Europe, Asia, and America have often asked me what will be our common future in view of my thesis of the Dark Ages and its implication for system transformation. In this final volume, I attempt to answer this by using the historical past as a template to anticipate our common historical future. It is based on a historical materialistic interpretation of the conditions during past Dark Ages and not on rational ideological/philosophical reasoning on what the future could be or should be. Hopefully, the reader will be satisfied with what I have attempted to do. What I have argued is that despite the connotation of Dark Ages and the social conditions that most of us are familiar with, these periods should be viewed as opportunities for innovation, experimentation, and social evolution. In other words, as my friend Matthias Gross has often remarked to me, I should label Dark Ages as Bright Ages!

My thesis on Dark Ages is also a theory of social evolution. It is a structural attempt to explain long-term social change by examining the relations between Nature and Culture, rather than just focusing on social relations as is so common in the humanities and the social sciences. It is my belief that perhaps the key toward understanding long-term system transformation lies in the study of relationship between Nature and Culture, rather than on socioeconomic and political relations. Hence the use of natural system triggers and tipping points in this volume. My three-volume contribution hopefully will spur others to pursue further this historical relationship between Nature and Culture.

For this final volume, I have benefited from the comments of Matthias Gross, Bill Devall, and Bill Thompson who have read the entire manuscript and provided their generous comments. Any omissions and errors are solely my responsibility. Furthermore, for the past decade my work on Dark Ages has gained much from comments and questions made by other friends and colleagues including the various audiences all over the world. In particular for this final volume, I wish to thank Ts'ui-jung Liu (Academia Sinica), Alvin So (Hong Kong University of Science and Technology), Kristian Kristiansen (University of Gothenburg), and Neil Roberts (Stanford University) for their hospitality and invitations to speak on the recurring Dark Ages and ecological futures at their respective institutions.

Many thanks are due to Marissa Marro and Elaine McGarraugh of Rowman & Littlefield for their efforts in the production process of this final volume.

Finally, to my wife, Elizabeth, despite her own demanding schedule as a resource teacher, offered to take care of our "boys" (Beau, Billy, and Zeus) including their daily walks so that I can spend a semester as a visiting scholar in Hong Kong at HKUST where this book was finally completed. I owe much to her for the optimism expressed in this book's conclusion.

Sing C. Chew
McKinleyville, California

Introduction

System Demise

SYSTEM DEMISE AND TRANSITION

The possibility of a global environmental crisis is looming on the horizon.[1] Excessive consumption, economic growth, and population explosion project a scenario of an ecological crisis of global proportions. This prognosis has been compounded further with the anticipation of two more destabilizing conditions: global warming and pandemic diseases. When global warming and pandemic diseases are added to the mix of an already precarious situation of resource depletion and ecological destruction, it further heightens the debate of the impending system crisis.

To date, this discourse on the impending system crisis has mostly focused on the future challenges facing the planet utilizing contemporary trends and tendencies of global warming and pandemic diseases (see Dessler and Parson 2006). What has been absent in this global dialogue—other than a few exceptions (Ruddiman 2006; Chew 2006c, 2007)—is a consideration of the data and information from the long-term past for our understanding of the projected challenges that humanity faces in the future. If the long-term past can be a guide, the past patterns of human choices and structural reconfigurations made during and after a prolonged period of crisis, transition, and perhaps devolution can, in my view, offer us a set of ideas of alternatives and hope for our common future.

From the long-term past as I have examined in *The Recurring Dark Ages*, there have been at least three recurrences (transitions) when the world system has been plunged into chaos and devolution. In our common history, these recurrences of demise have been labeled as the Dark Ages. To most of us, Dark Ages signify periods of great upheavals only in the socioeconomic and political realm. What is not as widely known is that these periods also undergo wide-scale evidence of ecological stress, natural disturbances, diseases, and climate changes. What is also interesting and seldom realized is that such historical patterns and characteristics of past Dark Ages seem to be *congruent with our present projections* of what the environment will be like in our common future.

Given such a congruency between past conditions and the projection of future trends, and if the available studies on global warming, resource depletion, and ecological degradation are analytically reliable, one should anticipate with some confidence that the mixture of projected destructive trends will generate pressure on the reproductive capacities of the system, and in a compounding manner, will most likely induce our world system into crisis and chaos in the future. If this should happen, it means that the conditions for *system reproduction* have reached limits and that the reproduction of the system within the existing structures is in peril. A period of transition will likely follow. We will again find the arrival of Dark Age conditions that I have explored in detail in *The Recurring Dark Ages*. What are these conditions? In *The Recurring Dark Ages*, I concluded that Dark Ages as recurring phenomena—with three periods (2200 B.C.–1700 B.C., 1200 B.C.–700 B.C., A.D. 300/400–A.D. 900) of Dark Ages occurring so far in world history—exhibit not only certain repetitive socioeconomic and political characteristics such as economic slowdowns and trade disruptions, political unrest and breakdowns, reduced social stratification and social simplification of lifestyles, deurbanization, increased migration, and population losses but are also periods of ecological changes and crises as exhibited by deforestation, soil erosion, and loss of biodiversity. Furthermore, associated with these socioeconomic, political, and ecological changes are climate shifts (temperature fluctuations and aridity), and the occurrences of natural disturbances such as volcanic eruptions and tectonic shifts.

It is to be noted that over world history ever since the collapse of the Roman Empire and the Dark Age of Antiquity (A.D. 300/400–A.D. 900) that followed, the world system has since then not experienced a Dark Age. Instead, there were several periods of continuous socioeconomic expansion and shorter socioeconomic downturns for the last one thousand years. Therefore, Dark Ages do not recur as frequently as the short business or economic cycles of expansion and stagnation such as the Kondratieff and Juglar cycles. As world history has informed us, Dark Age occurrences signal a systemic crisis of the world system. It is a period, as we have stated above, one of social chaos whereby the continued reproduction of the system has reached asymptotic limits.

What are these conditions and limits? In *The Recurring Dark Ages*, ecological stress (deforestation, loss of biodiversity, soil erosion), social and political unrests, urbanization, population, climate changes, diseases, and natural disturbances were considered as factors or conditions leading to system stress and devolution. Either as conditions or factors generating system crisis, these conditions have been examined intensively in my previous two volumes, *World Ecological Degradation* and *The Recurring Dark Ages*. For this final volume, I discuss the possibility of another Dark Age ahead and how climate changes and pandemic diseases might act as triggering points to the mixture of conditions and factors identified above, thus tipping an overstressed situation into system crisis, and perhaps transition.

The word *crisis* denoting the devolutionary conditions and tendencies of Dark Ages does not cover the range of reorganization and reconfiguration that goes on during this period. Despite the negative outcomes for human communities in terms of socioeconomic progress, this period induces patterns and tendencies that are different from that of a time of bountiful resources and optimum environmental conditions. Reconfigurations are made to meet the scarce ecological and natural resource conditions. From this, changes in the social system organization follow. Therefore, adaptations in the areas of energy saving, transformed lifestyles that consumed resources efficiently, and so forth, are promoted and tried. With such reduced socioeconomic activities that depict Dark Ages, Nature thus is given the opportunity to rest. From Nature's point of view, Dark Ages should not be feared. For those who are ecologically minded, the appearance of Dark Ages should be welcomed for they may lead to socioeconomic patterns that might be more innovative than the previous configurations, as they force us to make innovative choices under constrained conditions that otherwise would not have been taken. In this sense, Dark Ages should be seen as a *hopeful era*, as the degraded natural environment is given the opportunity to restore and rejuvenate itself, and the social system adapts innovatively to the changed circumstances. The system transformation that follows can be an ecologically progressive one.

CONCEPTIONS OF SYSTEM TRANSITION AND TRANSFORMATION

Explaining system transition and transformation has been one of the main preoccupations of social theory since its inception. From Emile Durkheim, Max Weber, and Karl Marx, the focus was on the transition from a traditional society to a modern day social formation with each system having its own distinct set of socioeconomic and political features. All three theorists agree that a system transition and transformation occurred from a traditional society to a modern (capitalist) society, though the manner on how the transition occurred and explained by them was quite different individually. Furthermore, in the case of Marx with his emancipatory intentions, a projection of a further transition to a socialist system was even made. In these early social theories, the overall emphasis was directed toward explicating the characteristics of the transformed society and identifying the features that set the newly formed social formation from the previous one in the areas of sociation (social relations), socioeconomic organizations, and political structures. There were lesser efforts on the part of Durkheim in identifying future system crisis, demise, and transition, for his theory was mainly trying to grapple with the changes occurring in the demise of the traditional society at that time.

Later social theorists such as Talcott Parsons (1971) continued along the same tack of Weber and Durkheim following the same transformative framework and reiterating the changes in socioeconomic organizations that occur in the shift from a traditional social formation to a modern society (liberal capitalist) with its distinctive features. With Parsons, the model of socioeconomic and political organization was couched within the framework of a social system in view of the fact that systems theory then was gaining increasing currency. Recalibrating Parsons's equilibrium model of a social system, Jürgen Habermas's (1975) theory of social formations was infused with immanent crisis and legitimation concerns. System crisis was a key element in his discussion of the dynamics of late capitalist social formations. However, the notion of transition and transformation beyond late capitalism was not introduced into his discussion of the system crisis of the 1970s. Other than a more general examination of the theory of progress and communication via a theory of social evolution, the articulation of system transition from the advanced capitalist system was quite silent in Habermas's work (1979).

This was not the case in world-systems analysis. In trying to grapple with the global crisis of the 1970s, the main proponents of this approach, Samir Amin, Giovanni Arrighi, Andre Gunder Frank, and Immanuel Wallerstein attempted to dissect the nature of the global crisis, and with each articulating a likely outcome (Amin et al. 1982). In this effort to understand the dynamics of the global system crisis, Wallerstein and the late Terence K. Hopkins even went further to project the demise of the capitalist world system and transition (Hopkins and Wallerstein 1996; Wallerstein 1998, 2004). For them, such a transition is in the process (Hopkins and Wallerstein 1996). Within such a purview, the capitalist world economy is viewed as a historical system with a beginning, a maturation, and an end. What this means to them is that when the trends of the world system proceed to such a stage when the cyclical rhythms can no longer restore the long-term equilibrium of the system, the end of the historical system is at hand. This end, according to Wallerstein (1998, 35), results in a system bifurcation and a transition to a different type of system(s): "we are living in the transition from our existing world-system, the capitalist world-economy, to another world-system or systems." While others were busy discussing the transition from socialist systems to capitalist systems in Eastern Europe and China in the 1990s following the collapse of the former Soviet Union, Wallerstein and Hopkins were instead suggesting a different transition—that of a total system shift to *something else*.

Exploration of such a system transition by Hopkins and Wallerstein (1996) theoretically finds little resonance in the later work of Andre Gunder Frank (1991) of the early 1990s.[2] By this time, Frank was questioning his own past writings, and Wallerstein's world-system approach which is that of a capitalist world system having emerged in Western Europe around A.D. 1500, and later expanding to encompass the globe (see Frank 1991, 1993, 1994, 1998; Frank and Gills 1992a, 1992b). Frank (1991, 1992a, 1992b, 1993) (with Barry Gills) had the view that the world system had a much longer history, perhaps as long as five thousand years. Besides noting Wallerstein's Eurocentric bias, to Frank (1998, 342–44), a discussion on world-system transition is problematic because there seems to be more continuity than ruptures in the evolution of the historical world system: "The argument here has been that historical continuity has been far more important than any and all discontinuities. Recognizing and analyzing this continuity will reveal much more than by myopically focusing on the alleged discontinuities." Listen further to how he (1991, 178) had expressed this earlier:

> the historical record suggests that *this same historical world economic and interstate system is at least five thousand years old.* There was *more continuity* than discontinuity or even transition of *this* world (capitalist) economy as an historical system across the supposed divide of the world around 1500 (italics in original).

Without delving into the debate on the merits of Frank's revisions of world-system analysis,[3] Frank's revised model tends to characterize world history as an endless cycle of repetitions of political, economic, and social processes and does not distinguish *the significance of historical epochs and the specificity of world-system transition and transformation.*[4] For Frank (1991, 1998), there had been no epoch-making transition and transformation of the world system since its emergence five thousand years ago, only shifting hegemonies from one region to another that have generated the various dynamics and trends of this world system. Taken along these lines of "denying" transition and historical epochs, Frank's revised model for understanding long-term change has the tendency to *flatten* world history and to make the history of the evolution of the world system (world history) devoid of critical ruptures and discontinuities. Therefore, the cyclical phases of economic expansion and contraction that underlie the dynamics of the world system for him are just *conjonctural* in nature and not *structurally* transformative.

Does the world system have transitional phases? I believe it has. I have noted previously of such periods since the world system's emergence five thousand years ago, and these transitions have signified the ruptures and discontinuities of the world system (Chew 2007). Furthermore, I believe that the present system is at its limits, and qualitative structural changes will follow in the evolution of this five-thousand-year-old world system. These ruptures and discontinuities to me historically are represented by the Dark Ages of world history for the last five thousand years. In a Braudelian sense, it is *l'histoire structurelle* of the world system. It covers the very long-term (*la longue durée*) with its structures and *conjonctures*, and you find occasions of such transition and transformation when the system has reached critical thresholds in its reproductive capacities.[5] For me, if the history of the evolution of the world system is depicted by *structuration*, *destructuration*, and *restructuration*, then the period of transition and transformation is characterized by the phases of destructuration and restructuration.

Outside the circle of world systems and world-system history approaches,[6] the issue of transition and structural changes has also been introduced in globalization style studies. Identifying such a transition phase is the 1996 information social system thesis of Manuel Castells. According to him, the present circumstances of global socioeconomic changes underline the destructuration and restructuration moments of the present transition and that these transformations have emerged in the form of a network society. Historically for him, toward the end of the last millennium, a new structure was being formed associated with a new mode of development—informationalism. A global restructuring process is ongoing shaped by technological innovations in electronic communication. The broad contour of this process is the emergence of an information society organized via networks. According to Castells (1996, 1997, 1998), such transformations have led to changes in cultural and labor practices along with power relationships. This transition to a new structure is a consequence of three independent processes: the information technology revolution, the socioeconomic crisis of capitalism and statism, and the rise of new cultural movements (Castells 1998). Sharing this view of global structural change (for us, transition) is Saskia Sassen's (2006) recent theoretical attempt to historically delineate the foundational changes that have taken place in the structures of the world system as a consequence of the globalization process. An epochal foundational transformation is underway that is restructuring the socioeconomic and political processes of the system and the generation of new global structures that is different from the past.

These various current attempts to define and distinguish global structures that are qualitatively global in nature and not simply just quantitative changes that have been going on for centuries assume that the system is in a phase of transformation to a set of different structures. If such is the case, the transition phase is under way, and perhaps a new epoch-making phase in the evolution of the world system is occurring. Whether this transformation (restructuration) process will continue uninterrupted is dependent on system adjustments, available social and natural resources, and the successful cooperation-cooptation of the different social movements in this historical process.

METHODOLOGICAL NOTE

Periods of transition therefore are important in our understanding of the social evolutionary processes of the historical world system. With a system crisis, a period of transition follows. A period of transition signals that the historical system is experiencing problems of reproduction. The oscillations of the system have become extremely chaotic as the systemic processes have reached reproductive and legitimizing limits, and continued maintenance and integration under the existing conditions become increasingly difficult. In other words, the cyclical rhythms of expansion and contraction are no longer capable of restoring long-term equilibrium and control. A transition follows that may last for long periods (centuries). If this is the case in theoretical terms, the long periods of transition that occurred in the past following system crises can reveal to us the level of complexity of the conditions and factors underlying the dynamics of the transitions, and as well, the structural reconfigurations that circumscribed the continued reproduction and social evolution of the historical system. An analysis of the historical conditions and circumstances can reveal the deep structural shifts that have taken place that underlie the continuities that persist. From this, one can then follow the trends and tendencies from such structural shifts in order to map the social evolutionary directions of the reconfigurations that have emerged through the adaptations that have occurred during the transition.

If one uses history as the basis for deciphering and delineating these structural shifts and trends, such a methodological tack will discipline one's efforts to map the future direction of the social evolutionary tendencies of the system following the period of transition, as it will force one to base our projections on past structural circumstances and limits, thus avoiding the tendency of projecting certain future directions that can be inspired and guided by moral or rational approaches.[7] Rather than relying on what happened historically and materialistically, these latter approaches tend to rely on the philosophy of progress in the domain of human affairs whereby freedom and equality are the premises of projecting the future trajectory of the world system(s).

My main intent in this final volume is to explore the possible trajectories that the world system might reconfigure during/following the transition (Dark Ages). This latter exploration, I will label as possible "ecological futures." Rather than identifying possible trajectories based on the "the exercise of our judgment as to the substantive rationality of alternative possible historical systems" (Wallerstein 1998, 1), the effort will be guided by the patterns of past structural changes and human choices made in the arenas of economic organizations, political configurations, and social-cultural lifestyles in previous transitions (Dark Ages) that I have discussed extensively in *The Recurring Dark Ages*. This means utilizing the past, and to try hermeneutically to anticipate the future by outlining the possible structural shifts that might take place during the period of transition using the structural shifts that have occurred in previous Dark Ages as a template. Methodologically, what I will attempt to do is to let "history be our teacher of life," and for it to identify for us possible "ecological futures" using the patterns of the past.

NOTES

1. See for example, The Club of Rome Reports (Meadows et al. 1972; Mesarovic and Pestel 1974; Meadows et al. 2004), The Brundtland Commission Report (World Commission on Environment and Development 1987), The Brandt Commission Report (1983), Global 2000 Report (U.S. Council on Environmental Quality 1980), UNCED Report (1992), IPCC on Climate Change (2001, 2007), *State of the World* (World Resources Institute 2005).

2. Or as the late Gunder Frank has referred to this period of his work as "Frank Mark 2."

3. This debate between Wallerstein and Frank on the nature of the world system has appeared in numerous places. For examples, see Frank 1991, Wallerstein 1991, Frank and Gills 1993, Frank 1998.

4. In our private conversation, Gunder indicated that there might have been some conditions in world history that have had transformative effects on the evolution of the world system. He mentioned two at least: the Neolithic Revolution and the Industrial Revolution that occurred in Sung China.

5. This conception is different from Hopkins and Wallerstein's (1996, 8) model whereby they see the transition as "a turning point so decisive that the system comes to an end and is replaced by one or more alternative successor systems. Such a 'crisis' is not a repeated (cyclical) event. It happens only once in the life of any system, and signals its historical coming to an end."

6. For a further exposition of the latter approach, see Denemark et al. (2000), Chase-Dunn and Anderson (2005), Gills and Thompson (2006).

7. For a different approach in projecting a future, see Wallerstein (1998). In assessing Wallerstein's approach, Anderson (2004, 70) states: "Whatever its merits, this is scarcely the end of utopia Marcuse had in mind."

1

The Conditions

Climate and Diseases

CLIMATE: PAST, PRESENT, AND FUTURE

In a postmodern world with time and space increasingly compressed because of various advances in technology, media, and production management (e.g., Castells 2000; Harvey 1990), the long-term past is seldom considered nor consulted. The past is conceived of being traditional, nonprogressive, and delimiting, for we live in the age of the present and future whereby they often melt together because of media innovations, the rapidity of transmission of information, and the flexible global production process. Such a conception of the past and the present is further bolstered by the philosophy of progress that permeates our knowledge base and worldview giving us an overall forward-looking confidence in shaping and determining the future. We thus have a forward-looking orientation of assuming that growth and development is inevitable. Hence, what is real is what we are experiencing presently and accessible immediately to us through the various media streams and knowledge bases. It is to the future that we gaze and not to the past that we turn toward.

In view of the above, our memories and knowledge of the past is sketchy and clouded by either the lack of interest or treated as a set of narratives that sometimes seemed provincial and determined by selected groups who fought hard to defend their rights and histories. The historical past then becomes a specialized area of study, and its role in the overall understanding of social reality is relegated to the specialists and kept restrained within the walls of universities and institutions of higher learning. Unfortunately, most of these scholars of the past have also accepted this delegation of the knowledge of the historical past and have not tried to overcome this determination of their craft. Moreover, they are also guided by the philosophy of progress and hence the past is a field of study, and not one that can guide or inform the modern mind. If this is the case, naturally the end result is that the layperson's understanding of the world is shaped by the knowledge of the present, for this is deemed important and should be treated as such by those that ruled or governed. Consequently, our worldview is shaped not by the practice of historical hermeneutics, instead the hermeneutics of the present predominates.[1]

Like everything else, such a worldview determined by the present has informed also the discourse on global climate change and its future trend. We are told of the potential increase of the global temperature in the future and its consequences for the reproduction of life. Because of the overwhelming focus on the *present*, we have developed the impression and opinion that the history of our climate systems has been devoid of chaotic change, and it is only due to the human impact on the environment for the last few centuries that we have somewhat changed the natural dynamics of the global climate for the future. The argument that usually follows is that we will witness climate changes in the future and that these changes are a consequence of human actions, whereas in the past if there were chaotic changes, the turbulent weather patterns of prehistory were deemed to be part of the natural cycles and that humans had no role in contributing to these changes.

It is clear that this conception of climate change based on scientific studies of the present and the immediate past has been accepted as the likely scenario for the future (IPCC 2001; Climate Science Report 2006). But what about the long-term past and the climate of prehistory: are the patterns similar or different to what has been projected for the future?

THE CLIMATE OF THE PAST

Unlike present climatic conditions, throughout the last ice age and prior to ten thousand years ago, Earth's climate was chaotic, exhibiting rapid fluctuations in snowfall and temperatures. This turbulent nature of the weather has been confirmed by ice core analysis from two international projects—Greenland Ice Core Project (GRIP) and Greenland Ice Sheet Project Two (GISP2)—undertaken in the early 1990s (Grootes, Striver, and White 1993). These chaotic fluctuations have been documented by other studies also undertaken in Greenland showing short-term fluctuations with annual average temperature rising and falling by up to 10°C, and snowfall declining or trebling by one-third over the course of a few years.

If we examine the long-term past it shows that the planet, after starting to warm up fifteen thousand years ago, saw the temperatures dropping again to near-glacial conditions two thousand years later. This drop in temperature that lasted for over a thousand years, categorized as the *Younger Dryas*, regarded by some as the last "shudder of the ice age," affected the entire Northern Hemisphere (Burroughs 2005, 45). The *Younger Dryas* was the sudden cold episode that occurred and interrupted the warming trend between the Last Glacial Maximum and the Holocene periods. The chaotic fluctuations came to an end three thousand years later (about ten thousand years ago) with the arrival of the Holocene period of stability in the climate pattern as Earth began to warm up again similar to the conditions fifteen thousand year ago when the warming started.

To climatologists, the Holocene is a period when the weather entered into a quiescent mode, and with the warmer temperatures the development of human communities followed (Burroughs 2005; Ruddiman 2005; Childe 1942). From the period known as the *Younger Dryas* to the Holocene (approximately ten thousand years ago), a transition occurred from a climate pattern that was chaotic to one of a benign weather; though the cooling did not end when the *Younger Dryas* period was over as there were other cooling periods until eight thousand years ago.

For some climatologists, the climate optimum only arrived around six thousand years ago. This was followed with the expansion of the forests in the Northern Hemisphere to the higher latitudes, though there were delays in Canada because of the slower disappearance of the Laurentide ice sheet. In other parts of the world, especially in the lower latitudes, the warmer temperatures led to stronger summer monsoons, thus increasing the level of rainfall in the tropics as well as much of the Sahara and the Middle East. During this period there was an absence of El Niño events, and the return of El Niño came about five hundred years later around 5,500 B.C. coastal Peruvian and Ecuadorian records support such return of these El Niño events (Rodbell, Seltzer, and Anderson 1999). From this, we should also realize that El Niño events occurred in the long-term past, and they are not a feature that has occurred only in the late twentieth century as we have normally assumed to be such a case without examining and realizing the climate patterns of the past.

The quieting down of the climate variability from the end of the *Younger Dryas* to the Holocene brought warmer and wetter weather, and such a climatic environment provided conducive conditions for the development of human communities as I have stated before. The Holocene period for the Northern Hemisphere is characterized by different periods of warmness and dryness. Between 11,500 and 9,000 years ago, the climate was warm and dry. From 9,000 to 6,000 years it was a warm and wet period, followed by a warm and dry period from 6,000 to 2,500 years ago.

The winters were about 1–3°C warmer and 2–4°C cooler in the Mediterranean region than the present time. Finally, from 2,500 years to the present the climate had changed to cool and wet conditions. Specifically, from 2,500 years to about 1,500 years ago, the conditions were warmer and drier, and then followed by a colder spell until A.D. 1000.

Of course, there are also regional differences in the Northern Hemisphere. It has been widely known that when winters are warm in Greenland, they tend to be severe in northern Europe and vice versa. Primarily this is due to what climate experts term *the North Atlantic Oscillation*. Not only does this weather pattern affect Europe, it also has an effect on the weather of the Middle East whereby a deep depression in the north near Iceland and a high pressure close to the Azores can produce westerly winds that will push mild air across Europe and the Urals while at the same time causing cold air over Greenland. The same flow also pulls cold air into the eastern Mediterranean and Middle East thus giving colder and wetter winters in this region. Besides the effect on Europe and the Middle East, such conditions also bring mild winters to North America. What this means is that northern Europe will have more snow while North America will experience less. As a result, this will affect the spring melt on both continents. Weather fluctuations thus are a result of this linkage through the North Atlantic Oscillation. Such tendencies mean that periods of intense circulation in the North Atlantic generate more difficult climatic conditions north of the Alps; however, those to the south of the Alps in regions like the Mediterranean, Anatolia, and the Middle East experience increased winter rainfall.

For the South Asian region, eight thousand years ago the climate was slightly warmer and moister than the present. The rainfall levels exceeded present values by 500 mm. This condition can be extended to northern Africa as well. The moister condition came to an end for the Saharan region, for by 5,500 years ago desiccation followed (DeMenocal et al. 2000). In Tropical Africa, this dry weather continued, leading to profound impact on the Nile floods about 4,200 years ago. Conditions in China around 5,000 years ago were much warmer and moister than at the present time, and cooling started around 4,000 to 3,000 years ago. Temperature changes in China, however, became more fluctuating over the last millennium with periods of warming and cooling take place.

It is clear that when the climate variability ended with the onset of the Holocene and stabilized around 6,000 years ago, there was a spurt of human development and ecological expansion. What this suggests is that limited climate variability provided optimum conditions for the natural and social systems to flourish and evolve (see Burroughs 2005). This benign weather pattern continued with the climate being warm and dry from 2,500 to 1,500 years ago. Overall, the climate trends were not as chaotic unlike the climate patterns of prehistory prior to 6,000 years ago that exhibited drastic shifts occurring over a very short period to the extent that such shifts could occur over a few years. Despite this seemingly benign weather pattern, it does not necessarily mean, however, that there were no long-term specific regional climate changes over the last 5,000 years. There were some changing trends that we need to note. As I have shown in *The Recurring Dark Ages* there were periods of warming and cooling during previous Dark Ages in Europe, China, the Near East, and the Mediterranean. Besides these earlier periods of temperature fluctuations starting 4,200 years ago, a warming pattern, the medieval warm period, started around A.D. 800 and lasted until A.D. 1200 when the climate turned cooler in the Northern Hemisphere. The Little Ice Age followed around A.D. 1590 and ended circa A.D. 1850 (Roberts 1998) or as long as between A.D. 1300–1900 according to Ruddiman (2005).

This brief excursus into the climate of the long-term past suggests to us that our conception of a climate that is not chaotic is really not representative of the long history of weather patterns. It is clear that prehistoric weather patterns were chaotic and variable, not benign and quiescent. The cessation of large-scale climate change at the end of the *Younger Dryas*, as well as less climate variability during the Holocene period, did usher in a more stable climatic period. However, even the Holocene, which is supposed to have quiescent weather though not as variable, has its periods of climate changes that are highlighted in the previous paragraph.

The current cries of global warming and the projection that we will be entering a new phase of changing weather conditions is alarming for anyone. But if such warnings are comprehended within the context that our weather has exhibited chaotic and variable conditions in the past should put us in the mind-set that these warnings should not be viewed as that unusual when juxtaposed within the history of our climate patterns. This is precisely what some pundits have argued in order to absolve humans from their responsibility in causing global warming. The issue here is not that the weather has not been chaotic in the past, but of our responsibility as human communities over world history in impacting the climate through our activities to reproduce our social systems. In other words, our activities have further added to the chaotic nature of our climate systems that in the long run has the potential of arresting social and biological evolutions. For some, the expected warming trend in the future dwarfs earlier natural variations over the last two thousand years.

As I have shown in *World Ecological Degradation* and *The Recurring Dark Ages*, past communities have had to face these climate changes and have either succumbed to the crisis conditions or have reorganized in order to resolve and transcend the barriers of socioeconomic reproduction that the changing weather incurred. It is this challenge that we again have to face in the future, and how we respond to the set of contingencies will reveal the commensurability of our actions to the actions taken by communities before us.

THE CLIMATE OF THE PRESENT AND FUTURE

The Intergovernmental Panel on Climate Change (IPCC) provides the most comprehensive statement of what climate experts consider the weather will be like in this century. It is this projection that we will use as the baseline in terms of assessing and projecting the changes of the twenty-first century. The IPCC (2001) predicts that human activities will lead to climate warming during this century. Global temperature will rise by 1.5–5.8°C between 1990 and 2100. In the most recent update, the *Fourth Assessment Report*, the IPCC (2007) again confirmed the projection by predicting a 66–90 percent probability that the best estimate for the low-rise scenario would be 1.8°C (likely range from 1.1 to 2.9°C) to 4.0°C (likely range from 2.4 to 6.4°C).

The projection of the IPCC has also been confirmed by the U.S. Climate Change Science Program (Thomas et al. 2006). Its first report indicated that the surface temperature and the troposphere have warmed in this century starting from 1958 onward. Global average temperature increased at the rate of about 0.12°C per decade since 1958, and about 0.16°C per decade since 1979. In the tropics, the increase was 0.11°C per decade since 1958, and about 0.13°C per decade since 1979. According to the U.S. Climate Science Program, the increase is a result of human influence on the climate system through greenhouse gases and aerosols, and the observed pattern cannot be explained by natural processes alone. This latest confirmation thus makes it a unanimous global agreement that the global climate is warming up. It is suggested that this rise started in the nineteenth century. If this projection is accurate, it would be the most significant and rapid shift since the start of the Holocene period.

The warming effect has the potential of increasing the frequency of El Niño events, and thus the increase in tropical storms and more damaging hurricanes. Furthermore, the global warming anticipated would mean a rise in surface temperature, leading to more water vapor added to the atmosphere resulting in heavier rainfall. Projections indicate the pattern of wetter winters and drier summers in the Northern Hemisphere and more intense monsoon periods in the tropics.

By no means is there agreement on this. For example, Burroughs (2005, 289) has argued that El Niño-Southern Oscillation warm events were less common during the warmest period of the Holocene between ten thousand years and five thousand years ago. Based on this historical record, he has predicted that global warming might not necessarily lead to more El Niño events and an increase in hurricanes. In the process, he (2005, 290) has stated that we will have a "reduced incidence of both warm and cool events, and hence less variation from year to year in the incidence of Atlantic hurricanes." This prediction of his is based on the assumption that the climate shifts we are going into are similar to those that occurred in the Holocene. However, if humans have brought forth changes to the global climate that could lead to sudden changes in the climate, then we will be facing a set of different circumstances whereby the global warming "could lead to a shift in the thermohaline circulation in the North Atlantic" resulting in different weather patterns (Burroughs 2005, 290).

It is clear that, according to the IPCC (2001, 2007) and the U.S. Climate Science Program, humans have caused climate change (in terms of global warming and other outcomes) that will exacerbate during the twenty-first century. The beginning of human impact on the weather has commonly been considered a recent phenomenon following the exponential increase in greenhouse gases and aerosols emissions following runaway economic growth and deforestation. It is widely thought and assumed that this impact started about one hundred to two hundred years ago with the advent of the Industrial Revolution in the 1800s. We find this in the works of scholars such as Burroughs (2005), who have accepted that human-generated impacts on the climate patterns have been of recent memory whereas in the long-term past the weather changes were basically outcomes of natural cycles.

This common point of view shared by most climatologists, however, needs to be reconsidered in view of the calculations made by Ruddiman (2001, 2003, 2005). Ruddiman's hypothesis is interesting for he postulates the human impact on the climate has deeper roots for it started eight thousand years ago, and thus has not been caused by humans only recently. For him (2003, 2005), global warming started as soon as humans conducted farming activities. Thus, human impact on the climate started with the Neolithic Revolution and with the advent of agriculture in the Fertile Crescent of Mesopotamia and the Yellow River Valley of northern China where land clearing and deforestation were the sources for the increase in carbon dioxide and methane. Despite the low rate of increase in carbon dioxide and methane concentrations, according to Ruddiman it is the slow increase spread over at least eight thousand years for carbon dioxide, and five thousand years for methane that is responsible for global warming. This warming is not a product of natural cycles of the climate system. Specifically, deforestation to increase clearings for cropland in China, India, and Europe provided the increase in carbon dioxide, whereas methane had several sources: rice farming in the tropics (China and Southeast Asia), animal husbandry throughout Europe, China, and India, and biomass burning and human wastes. However, the most significant factor for the rise of methane according to Ruddiman is farming through irrigation.

If what Ruddiman (2005) has asserted above is a plausible explanation for global warming, then it means that human impact on the climate system has had a very longer-term origin than what is commonly accepted and assumed. By no means does such a thesis negate the anticipated global warming that has been projected for the twenty-first century by the IPCC (2001, 2007) and the U.S. Climate Science Program. Rather, such a thesis underlined the long-term impact on the environment, in this case, particularly the climate system by human communities. It suggests that human-induced global warming has a longer history than what is commonly accepted—a phenomenon caused by the Industrial Revolution two hundred years ago.

Besides identifying the anthropogenic sources for global warming that cannot be the result of natural cycles, Ruddiman (2005) has also asserted that the appearance of diseases during the last two thousand years has also impacted the climate, in this case a cooling effect. The logic is as follows: diseases lead to population decreases that concomitantly result in fewer human activities—such as deforestation—that would mean fewer carbon dioxide emissions, thus ultimately generating short-term climatic cooling instead of warming. These decreases tend to cluster during Dark Ages. There were three carbon dioxide minima for the last two thousand years: one between A.D. 200 and A.D. 600 (the Dark Age of Antiquity), a short drop between A.D. 1300–A.D. 1400, and a final one between A.D. 1500 and A.D. 1750. Given such logic, the Little Ice Age interval between A.D. 1300 and A.D. 1900 according to Ruddiman (2005) was a result of such a connected series of events. If we agree with this, diseases are now added into the mix of factors that impact climate change, except that in this context, a lesser human impact means the tendency toward a cooler climate following a drop in carbon dioxide concentrations.

If human-induced global warming has had a long history, then the patterns of the past in terms of trends and structures can provide us with useful vantage points whereby we can better anticipate future tendencies of the evolution of the system in an era of systemic crisis, and realize that what is considered "new" in terms of development of structures and trends might not necessarily be the case.

DISEASES

When humans were hunter-gatherers and widely scattered with a low population level, they were relatively free of major communicable diseases. This condition was to change with the advent of agriculture and pastoralism when humans become more numerous in numbers, clustered in higher density urbanized environments, and interrelated with Nature and domesticated animals, thus making them susceptible to infectious diseases (Bray 1996; McNeill 1975). Undertaking agricultural cultivation and rearing domesticated animals required a sedentary lifestyle. Such practices enhanced the possibility of contracting diseases, because remaining in one place provided the possibility of coming into contact with feces of other humans and animals, and from these infections came the possibility of transmission of diseases from animals to humans and humans to humans. Where human communities remained in one location on a permanent basis and had to rely on a local water source, contaminated water supplies seemed more likely to occur and hence the transmission of communicable water-borne diseases. Irrigated agriculture also provided opportunities for disease parasites to be transmitted; for example, ancient Egyptian irrigators suffered from infection as early as 1200 B.C. (McNeill 1975).

As human populations increased and the density of sedentary communities also expanded, infections were transferred repeatedly as a result of the increased contact between humans and also between animals and humans. As such, urbanization has the effect of heightening the transmission rates because of such circumstances. Urbanization meant a higher division of labor, and with this the surplus generated was exchanged locally or through trade. The latter not only allowed the support of a larger population, but also provided the conditions for movement of peoples between one urban center and another and between regions. Ships and caravans were the modes of transportation used to ferry the goods. Such trading encounters between groups of merchants and exchange of products including food stuffs were also avenues whereby diseases could be transmitted or communicated. Furthermore, the mode of transportation such as trading ships has been known to carry disease-infected animals, such as rodents. As urbanization continued, urban dwellers, having developed certain immunity to childhood diseases, would through travels or conquest spread diseases to communities that were isolated and had not been exposed to such diseases.

Over world history, these various avenues and configurations have determined the dissemination routes whereby communicable diseases have spread. The histories of epidemics and pandemics of diseases have been documented in several studies, and for the last two thousand years they tended to occur in clusters and during periods of environmental change and socioeconomic crises (McNeill 1975; Bray 1996; Ruddiman 2005). The types of diseases that have occurred over world history are smallpox, bubonic plague, typhus, malaria, influenza, and cholera. Ruddiman (2005, 132) has categorized and collated the epidemics and pandemics of diseases by region for the last two thousand years (see table 1.1). Geographically, these epidemics and pandemics have occurred in the continents of Asia, Africa, Europe, and the Americas. The simultaneous occurrence of these diseases during a certain historical period underscores the level of connectivity between these regions. This connectivity is determined by the trading regimes and the series of wars and conquests that took place over the last two thousand years. The clustering of disease occurrences especially between A.D. 200 and 600 during a period of climate change and socioeconomic crisis should be noticed, and especially when it was during the Dark Ages of Antiquity.

During the Bronze Age, accounts of epidemics have been noted even as early as 1500 B.C. in the Old Testament, when the plague was reportedly

Table 1.1. Epidemics and Pandemics of Diseases by Region

Year	*Region*	*Disease*	*Intensity*
79–125	Rome	Malaria	Local Epidemic
160–189	Roman Empire	Small Pox	Regional Epidemic
265–313	China	Small Pox	Regional Epidemic
251–539	Roman Empire	Small Pox/ Bubonic Plague	Regional Epidemic
540–590	Europe, Arabia, and North Africa	Bubonic Plague	Pandemic
581	India	Small Pox	Regional Epidemic
627–717	Middle East	Bubonic Plague	Regional Epidemic
664	Europe	Bubonic Plague	Regional Epidemic
680	Med. Europe	Bubonic Plague	Regional Epidemic
746–748	Eastern Mediterranean	Bubonic Plague	Local Epidemic
980	India	Small Pox	Regional Epidemic
1257–1259	Europe	Unknown	Regional Epidemic
1345–1400	Europe	Bubonic Plague	Major Pandemic
1400–1720	Europe, North Africa	Bubonic Plague	Regional Epidemic
1500–1800	Europe, Americas	Small Pox	Regional Epidemic/ Major Pandemic
1489–1850	Europe	Typhus	Regional Epidemic
1503–1817	India	Cholera	Local Epidemic
1817–1902	India, China, Europe	Cholera	Pandemic
1823–1889	Europe	Influenza	Regional Epidemic
1918–1919	Global	Influenza	Pandemic
1894–1920	Southeast Asia	Bubonic Plague	Regional Epidemic

Source: Ruddiman 2005, 132.

afflicting the Egyptians; and the Hebrews were promised that the sickness would be taken away when they returned to their homeland, which was a drier and sparsely settled colony away from the humid, crowded cities of Egypt. Such conditions we know are less likely for diseases to spread. A millennium later, around 430 B.C., the plague occurred in Greece. From its origins in Ethiopia, it spread northward to Egypt and Libya, and with the trading connections between North Africa and mainland Greece, the disease arrived by ship from Egypt.

The foci of infections concentrated in Mediterranean Europe, Egypt, Mesopotamia, India, and China through war, trade, and travel. By the Roman period, with the trading connections firmly established between eastern Mediterranean, India, and China, the occurrence of diseases became increasingly regionalized. Besides plague, malaria also occurred predominantly during the Roman period, with accounts of its appearance as early as A.D. 79. Its origins are believed to be from the rainforests of Africa where the disease traveled up the Nile to the Mediterranean and eastward to Mesopotamia. The infection was spread throughout Europe by Roman soldiers in the many Roman war campaigns. The scope extended as far north as Denmark and England. Besides wars, migrating groups whose movements were a consequence of changing climate also ensured the spread of diseases. The Huns from their homeland in Central Asia migrated through the steppes and continued westward to Europe bringing the plague with them. In turn, the Huns also came into contact with diseases that they were not immune to and thus were infected as well (Bray 1996).

Outbreaks of plague occurred between A.D. 542 and A.D. 748. A major epidemic hit the Roman Empire to the extent that Constantinople was losing five thousand persons a day to the disease in the first three months of the outbreak in A.D. 542 (Bray 1996). Other areas where the epidemic was prevalent were Upper Egypt, Nubia, Europe, Syria, and Persia. Recurrences of the plague appeared in the Middle East around A.D. 627–717, and in China around A.D. 610 (Karlen 1995). This pattern continued with its appearance in A.D. 743 in Syria, moving west to Egypt and eventually to Greece and Constantinople.

Following the end of the above epidemic, there was a lull from A.D. 900 onward. With the weather changing to mild winters and dry summers farming resumed. The forests regenerated as a result of the lowered population following the various epidemics that afflicted Europe, the Mediterranean, and the Middle East. Recovery allowed Europe's population to double between A.D. 900 and A.D. 1300. However, by this time also, the climate had started to change again. The temperature had started to turn cold and wet with historians claiming this to be Europe's Little Ice Age. Heavy rains, crop failures, and famine followed, and it lasted from

A.D. 1309 to 1325. These conditions were not only limited to Europe. In China a similar situation persisted. In A.D. 1333, China experienced drought, famine, earthquakes, and epidemics. In Mongolia, the weather changes forced the nomadic tribes such as the Mongols to migrate to new territories in search of food and pasture. From these areas in Central Asia, the plague was carried via rodents and the Mongol horsemen across Central Asia to Europe. Besides this vector of transmission, the disease was also brought to Europe by trading ships and caravan routes from Central Asia and the Middle East. By the 1330s, the plague had also spread to China, India, and the Middle East. It also covered most of southern Europe that had trading relationships with the Middle East and the eastern portion of the Mediterranean. Population losses were high. Europe was supposed to have lost 25 million of its population (Bray 1996). The Islamic world had population losses between one-third to one-half of its people, and India experienced the same level of losses. China lost half its population (65 million) between A.D. 1200 and 1400.

By no means were the plague epidemics that extended to a pandemic from A.D. 1345–1400 the end of this virulent disease. There were decadal repetitions in Europe and North Africa between A.D. 1400 and A.D. 1720. Added to this deadly disease was smallpox, which had its occurrences earlier in the Roman Empire and China from A.D. 160–189 and A.D. 265–313 respectively, and became a pandemic between A.D. 1500 and 1800 in the Americas and Europe. For the Americas, the disease was brought to the New World by European colonizers. The Spanish conquest of the West Indies led to a major epidemic in A.D. 1518 that spread to Aztecs in the Yucatan following colonial penetration by Cortés. By A.D. 1530, small pox had raced throughout the Americas from the Great Lakes to the Pampas. The Mayans were also afflicted, and it followed south to the Inca Empire. Small pox was also prevalent along the Mississippi River valley, and from 1539 to 1542 was endemic in this area. Besides small pox, measles also accompanied the Europeans and made its appearance in Cuba, Honduras, Mexico, and Central America.

Population losses were high from these disease outbreaks. Following Cortés's arrival in Mexico, fifty years later the total population went from 25 million–30 million persons to about 3 million (Bray 1996). In another estimate, in the sixteenth century, smallpox killed about 18 million people in Mexico, and in Peru, the Incas were reduced from 8 million to 1 million (Karlen 1995). On an overall basis, some estimates suggest that the New World population suffered losses from diseases totaling about 90 percent following the arrival of the Europeans.

Two other virulent diseases that impacted Europe and Asia were typhus and cholera. Typhus appeared in southern Spain in the 1490s, and from this region it spread into France. It continued to be around from the late fifteenth century till the mid-nineteenth century and was of epidemic proportions in Europe during certain periods such as the Thirty Years War (A.D. 1618 to 1648). It also appeared in the urban environments of Europe especially during the Seven Years War and the French Revolution. In the early parts of the nineteenth century, typhus was prevalent among the armies of Napoleon during his Russian campaign. It continued its ravages in Europe throughout the nineteenth century even through to the World War I in early twentieth century.

Cholera, a waterborne infection, appeared in Europe following its appearance in India from A.D. 1503 onward. A local epidemic appeared in India during the sixteenth century with outbreaks occurring through to A.D. 1817. It has been suggested that it could also have spread to the coastal regions of China by religious pilgrims and trading ships. With trade and troop movements, cholera managed to be transmitted to other distant shores. It entered Afghanistan, Persia, and southern Russia. From Russia it moved to Poland, Germany, and finally reached England by ship. The start of a pandemic from A.D. 1817 onward to the beginning of the twentieth century saw the disease afflicting the various populations of India, China, and Europe. From the 1830s onward, every part of Europe was impacted as far north as the Scandinavian countries. From England, it was carried to Ireland, and with the Irish immigrants cholera was exported to the United States. From here, with trade and travel it moved to Mexico and Cuba.

Besides cholera, influenza was also a disease that reached epidemic and pandemic proportions. The earliest account of its existence is around A.D. 1387, though some, such as Bray (1996), have documented its existence as early as 412 B.C. The first pandemic is supposed to have occurred in the sixteenth century in Europe. Its origin supposedly began in Russia and Central Asia. In Europe the epidemics of influenza occurred over the course of the fourteenth century to the nineteenth century. In 1889 an epidemic of the disease killed about a quarter of a million people in Europe. Over world history the extent of influenza pandemics stretched over the globe with the pandemic of 1918–1919 having one of the highest mortality rates. It is estimated that the pandemic killed about 22 million persons. The United States suffered a loss of about half a million deaths, while the United Kingdom lost about 200,000 of its population. India, which was greatly affected, lost about 12 million persons. In Africa, there were over 500,000 dead in the Belgian Congo alone.

Disease occurrences tend to cluster during certain periods of the last five thousand years of world history. Table 1.1 has shown that during the most recent Dark Ages, that of Antiquity (A.D. 200/300–A.D. 900), there were numerous regional epidemics and pandemics. As I have discussed in *The Recurring Dark Ages*, changing weather during Dark Ages can cause disease outbreaks, especially those that are transmitted by animals. Wetter climates can increase the animal population, such as rodents, leading to certain circumstances of animal mobility that often results in a higher frequency of contacts between animal and humans, and thus a higher rate of infections. Besides changing weather patterns, trade, travel, and wars also lead to further transmissions. The propensity of human communities to increase population and the incessant drive toward urbanization and higher conglomerations exacerbate the transmission of diseases. Inequality between groups and the disparate conditions of living environments coupled with poor nutrition among the dispossessed ensures the continued propagation of diseases. During the past Dark Ages, these trends and tendencies reached asymptotic levels, and the prevalence of disease epidemics and pandemics seemed to have the capacity to reduce or impact global population levels. What this engenders is a reduction in utilization of natural resources and deforestation and a lull in carbon dioxide and methane emissions as a consequence of a lowered population. Such a line of reasoning follows that of Ruddiman's (2005) where he postulated that lowered carbon dioxide emissions during certain periods is a consequence of diseases causing reduction of the human population, and hence a lowered level of economic activity resulting in less carbon dioxide emissions.

Besides global warming and the occurrences of pandemics occurring in a period of uncertainty and change, we find that socioeconomic and political configurations are under stress with cultural practices and lifestyles readdressed in light of changing conditions. These moments of transition whereby the old frameworks and structures are being reformed in relation to the conditions, the emergence of alternate worldviews and life practices that shape socioeconomic and cultural life will condition the transformation of the system as it continues to evolve and adapt to changing circumstances and environmental conditions. Throughout world history, as I have expressed in *The Recurring Dark Ages*, we find the appearance of such institutions. During the late Bronze Age crisis, for example, we find the development of the "poleis" as a political framework to organize the political economy of Greece. In the Dark Age of Antiquity, another institution, monasticism, emerged to address the socioeconomic and ecological crisis then. Following its rise in Egypt in fourth century A.D., and its successful establishment as part of the religious and institutional structures in the Roman Empire and Europe, the monastic movement with its social, economic, and religious transformative characteristics provides an example of an institutional mechanism that evolved during the Dark Age of Antiquity, and in turn shaped system evolution and transformation. The appearance of alternate worldviews and life practices seem to recur through world history, and especially during phases of system crisis. In the late twentieth century, this again is the case with the call for bioregionalism as a response to the global socioeconomic and ecological crisis. It is to these alternate worldviews and life practices that we turn to examine in order for us to better anticipate the future trajectory of system transformation and adaptation.

NOTE

1. On historical hermeneutics, see Gadamer (1975).

2

The Reactions

Alternate Life Practices

Times of uncertainty and distress generate reassessment and reorientation. Over time, social institutions emerged to address these periods of turbulence and uncertainty. As expressed in *The Recurring Dark Ages*, the crisis of Antiquity (A.D. 300–900) was one such period when the socioeconomic and political disruptions and chaos led to certain transformations that, in the long run, were determining factors in the development of the social formation in Europe after the end of the Dark Ages in the tenth century. The emergence of Christian monasticism in the fourth century A.D in Roman Egypt was one such transformation. From its spread throughout the Roman Empire and Western Europe from the fourth century until the twelfth century, Christian monasticism was one of the most continuous forces that played a "determining factor of the most fundamental kind in the formation of medieval civilization" (Cantor 1994, 146; Daniel-Rops 2001).

The rise of Christian monasticism with the appearance of spiritual leaders, individuals, and communities separated from the rest of the world, and framed within the Christian tradition, is a phenomenon that needs to be understood and evaluated in light of the Dark Age conditions of the times. Christian monasticism, operating within a religious framework that requires retreat from the inhabited urbanized communities of the times and seeking the return to one's original state, was a development that was not commensurate with the cultural and institutional religious practices of the period. Therefore, it can be viewed as a movement offering an alternate worldview and a set of religious and social practices that was detached and religiously different from the ongoing practices of the times. In this light, Christian monasticism can be perceived as a reaction to the social, economic, political, and religious conditions of the day, and one that offers relief and salvation from the turbulent conditions of the Dark Age of Antiquity. It can also be conceived as a social institution offering an alternate worldview and lifestyle choices for the reconfiguration of the social formation during the medieval period. Especially during the turbulent periods of the Dark Ages, Christian monasticism provided the framework that was quite different from existing conceptualization of socioeconomic and religious life and practices.

Similarly, in our current era of crisis that encompasses ecological, socioeconomic, and political dimensions almost 1,700 years later, the present tumultuous conditions have also sparked movements expounding alternate worldviews and lifestyle options. One such movement is bioregionalism with its emphasis on localism and proposing a life practice centered around localism and self-sufficiency that is very different to contemporary accepted legitimate land use, ethics, and practices. In fact, the bioregional (life place) vision is a direct contrast to our contemporary worldview that underscores the themes of globalization, technologization of life, and hyper consumption. Therefore, along a similar vein to early monasticism's reaction to institutionalized religion then, bioregionalism as a life practice also plays a similar role in the contemporary crisis era, for it offers alternatives that are the opposites of what have been institutionally accepted.

Given such a parallel development in spite of the fact that the rise of Christian monasticism over 1,700 years ago as a consequence of a long period of crisis, can an evaluation of its emergence, growth, and perpetuation inform our understanding of the patterns and tendencies existing in our present conditions of crisis? I believe it can, and by considering monasticism and bioregionalism as outcomes to crisis conditions of the system, we can better understand and project what we face in the future in terms of system transformation and organization.

THE RISE OF CHRISTIAN MONASTICISM

There are various interpretations and explanations for the emergence of Christian monasticism. With its origins in Egypt starting around the fourth century A.D., some have viewed monasticism's appearance as a reaction to institutional Christianity and the imperial adoption of Christianity in the fourth century A.D. (Workman 1913; Griggs 1990); while others have attributed its emergence to a call for a return to the values of Christian martyrdom (Malone 1950). Whatever the various reasons provided, the beginnings of monasticism can also be explained as an outcome to a structural transformation that started in the third century A.D. in terms of socioeconomic upheavals, political distress, and ecological crisis (Frend 1972; Griggs 1990). This period of structural transformation is commonly understood as the Dark Age of Antiquity, and it was also the start of the decline of the Roman Empire that dominated Europe and the Mediterranean. It was a time of widespread anxiety with changing socioeconomic values and conditions. According to Dodds (1965), religious feelings and reactions then were very intense, especially during a period when material decline was the steepest.

In the Roman colonies, these reactions were fueled further by the tax imposed by imperial Rome. As a result, we find attempts by the local citizenry to escape from Rome's tax burden. One common reaction was a flight to the monasteries by farmers so as to avoid tax obligations (Frend 1972; Daniel-Rops 2001). Historical evidence shows this occurring in Egypt and other parts of the Roman Empire such as Palestine and Syria. Interestingly, it is also in Egypt and Syria that we witness the rise of monasticism.

By the fourth century in Egypt and Palestine there were organized monastic communities living in the deserts. Early Christian monasticism's rise in Egypt in the fourth century A.D. gained momentum as it spread throughout the Mediterranean and Western Europe (Griggs 1990; Daniel-Rops 2001). By the A.D. 340s, it was beginning to be felt in the West with the establishment of monasteries in Italy. In addition to taxes, political unrests also fueled the participation of the populace in the monastic order. The upheavals in the mid-fifth century in Gaul led to the retreat of local aristocrats to the monasteries leading to the further propagation of the monastic ideal. In Britain and Ireland, there is evidence of different monasteries being established by the end of the fifth and the early sixth centuries A.D. In Byzantium, by the fifth century A.D., Christian monasticism interacted in harmony with the eastern Roman Empire.

The monastic mission was one of the renunciations of personal property and the quest for spiritual perfection. The former meant the abandonment of worldly goods and through contemplation achieve union with God. This required prayers and the pursuit of a life of solitude. Such intentions for perfection can be obtained through a withdrawal from society. It meant moving away from urban communities and living away from others. This practice also required the rejection of materialistic life and urbanized living. Because of monasticism's origins in Egypt and Palestine, the desert was the natural place of refuge from civilization. Here, the monk can seek solitude and salvation and live an ascetic experience.

Accordingly, true knowledge (*gnosis*) is to return to one's original state, and after achieving this one can aspire eventually to have a union with

God. The acknowledged leader of this mission in early monasticism was Antony, an Egyptian (Meyer 1950; Athanasius 1980). Early monasticism pursued two organizational paths. One was living a hermetic life in groups in the desert while the other was toward the establishment of a communal organization or monastery. This latter initiative was led by another monk, Pachomius, who established monastic structures of this nature (Dunn 2000; Lawrence 1984).

According to St Athanasius early monastic life started with Antony giving up his comfortable life after hearing the Lord telling him "if thou wilt be perfect, go sell all that thou hast, and give it to the poor, and come, follow me and thou shalt have treasure in heaven" (Meyer 1950, 19). Following the dictates of the Lord to relinquish his personal property, Antony moved to the desert to find solitude and solace. On another level, this rejection of urban life to live in the desert suggests a reaction and disavowal of civilization's social organizational principle of urban living and a return to Nature. Urbanization has severe consequences on the ecology of the land as I discussed in *World Ecological Degradation*. To this extent Frend's (1972) suggestion that perhaps this act of Antony of selling his farm of 300 *ąrourae* (an *aroura* is about one-third of a hectare) was not just only following God's message for there were other reasons to be explored. The land of the farm was in a poor state, and by the sale Antony was also relieving himself of the burden of having to pay the taxes (Frend 1972, 11).[1] As Frend (1972, 11) has further indicated, even Antony's biographer, Athanasius, has written that with the sale the grumbling of the tax collector was no longer heard.

Again, on another level beyond the religious, this act of Antony could also have ecological and economic roots. Placed within a wider structural context, this sale of the farm to end the obligation of paying taxes along with the poor state of the land alerts us to two issues affecting Roman Egypt: the ecological degradation of the landscape because of the pressure to undertake excessive agriculture practices so that taxes can be paid, and the tax obligations imposed on the farming population.[2] Changing weather patterns resulting in irregular Nile flows added further to the crisis conditions. By the late third to the fourth centuries, even among the monks there were conversations about the flow of the Nile (Brown 1978). This set of circumstances heightened the social tension among the populace (Green 1986). The tension of living in the urbanized world proved to be unbearable for Egyptian farmers and proprietors. Therefore, we get the impression that for some, the move to the desert to become monks was driven by a crisis in needing to meet taxation obligations, land deterioration, and the changing weather. The end result was disengagement or *anachōrēsis* on the part of the peasants, farmers, and proprietors from Egyptian society. Placed within such a context, the emergence of monasticism in

Egypt can be attributed to a set of structural conditions during a period of crisis known as the Dark Ages.

MONASTIC PRACTICES AND ORGANIZATIONAL FRAMEWORKS

Imbued with the need to live an ascetic life, the move to the desert was to provide the opportunity to have solitude and to be able to seek salvation away from other people and urbanized life. The city or urbanized settlements were considered places of temptation and evil. Furthermore, it was deemed a place determined by traditional rules and obligations. Given this, the desert or wilderness then became the anticity, where the monk was relieved of temptations and was able to live an existence of complete autarky and self-sufficiency. Desert living was a step toward one's salvation for the person would be like what she or he started at the beginning of life on earth and living in unity with Nature (Bergmann 1985). By living in such a proximity to the natural state, the monks were also practicing a life of self-sufficiency. Antony, according to Athanasius (1980), was living such a life after moving to the desert. With the hoe, axe, and some grain provided by some of his brethren when they visited him, Antony proceeded to till a small patch of land and sowed the seeds and subsisted on his own:

> He thought the matter over and asked some of those who visited him to bring him a two-pronged hoe, an axe, and some grain. When these were brought, he went over the ground about the mountain, and finding a small patch that was suitable, and with a generous supply of water available from the spring, he tilled and sowed it. This he did every year and it furnished him his bread. He was happy that he should not have to trouble anyone for this and that in all things he kept himself from being a burden. (Athanasius 1980, 63)

Living self-sufficiently meant also living with the contours of the land. In terms of relationships with other living things such as animals, we obtain the impression that the monks were living in harmony with them, and this is not because of their nonmeat diet. Antony's experiences with wild animals that visited his small garden give us an image of this relationship:

> At first wild animals in the desert coming for water often would damage the beds in his garden. But he caught one of the animals, held it gently, and said to them all: "Why do you do harm to me when I harm none of you? Go away, and in the Lord's name do not come near these things again!" (Athanasius 1980, 63)

In terms of diet, what was consumed was supposed to be natural and what was available such as water, salt, roots, and herbs. Of course, meat was taboo (Dunn 2000). In terms of diet, Syrian monks went to the most extreme. They subsisted on water and grass (Bergmann 1985). This daily intake was to change over time as the monasteries developed different organizational forms following the admission of different social classes joining the monasteries. The drinking of wine was allowed, and other types of food beyond roots and herbs were added including lentils, salted olives, grapes, prunes, and salt fish.

By the time of Antony's death in A.D. 356, there were desert communities in the mountain of Nitria to the west of the Nile Delta, and the inner desert of Scetis. Similar settlements also appeared in Syria and Palestine. In Syria, some communities were living in caves. It is estimated that these colonies had population of monks of several hundreds. Cave living was chosen as it would bring the monks closer to a natural state (Bergmann 1985). The organizational feature of these settlements was comprised of a complex of buildings including bakeries, a church, and accommodation for guests who visited. The monks produced baskets, mats, ropes, and linen for their own needs, and in some cases they were exchanged with the local community for other necessities. This exchange was handled by agents from the local villages.

Organizationally, at this early stage life was not structured by rules. The only communal activity was the gathering for prayer, and the abbot provided counsel to those who requested it. He also provided guidance to the neophytes. Living in such an eremitical solitary environment was not the only organizational form that existed.[3] Others followed that of the coenobitical monastery (Lawrence 1984; Goehring 1990; Daniel-Rops 2001). Influenced by Pachomius, a Coptic Christian, it was a communal monastery. Established first at Tabennesis replacing a deserted village in the Upper Nile, this monastery flourished and more were established as time progressed (Goehring 1990). The typical monastery of this type had a community of about one thousand monks. Surrounded by walls, it had a wide precinct with buildings such as a church, common refectory, and a hospital. There was also a building for visiting guests. The dormitories or houses of the monks were laid out like a Roman legionary camp as Pachomius was influenced by his experiences in the Roman legion. Each house had twenty monks living in it with a prior as its head. Here, we see the emergence of stratification that was followed with the establishment of monastic rules. These rules focused on the need to surrender the personal will in favor of a superior. Notwithstanding the stratification process that was taking place, there was, however, communal fellowship exemplified by the common meal and the common prayer.

In the Pachomian monastery, work was considered a virtue, and, especially with most of the monks coming from peasant origin, there was little provision for intellectual activities. Production of linen, ropes, mats, and baskets was undertaken in the monastery from materials gathered locally. These monasteries after they have grown in number and size had considerable economic impact on the local community. By the sixth century A.D., within the vicinity of Alexandria there were about six hundred monasteries with about thirty-two associated farms (Hardy 1952; Frend 1984). The production of various goods required venturing to the local areas for raw materials. Rushes were gathered from the neighboring river system for the plaiting of ropes and baskets. Agricultural production was undertaken along with the growing of fruits. Shearing of goats for shirts was also undertaken in the monastery. Other monasteries were more diverse in terms of the goods produced. Beyond what has been described, books, leather goods, shoes, wooden products, and metal goods were also manufactured for exchange (Dunn 2000). As established institutions, detailed financial records were kept. The provision of other services to the local community was also undertaken and reflected in the regulations for monks in terms of their responsibilities (Goehring 1990). The economic activities of the monasteries were under the direction of a housemaster. The manufactured items of the monasteries were transported down the Nile for exchange as far as Alexandria, and in the early periods this was under the management of a steward.

With the founding of monasteries in the rural areas of Egypt and their associated economic activities, the impact of these monasteries in terms of revitalizing abandoned farms was quite significant. Over time, more land was added to the monasteries as well (Dunn 2000). The growing social and economic interdependence between rural communities and the monasteries became firmly established. The leaders of these monasteries were the new religious men of antiquity with the monastery as the new base of social and economic power (Goehring 1990). During the Roman period, especially during turbulent times, the monasteries became the interface between the people and Roman administration. In other words, they acted as the buffer between the state and its subjects. At other times, they served as refuge for the surrounding population during periods of political distress. Food and medical services were provided during wars and raids.

By the sixth century, the monasteries were much more diverse and became more institutionalized than what they had been in the fourth century A.D. (Dunn 2000). Their presence was being felt all over Europe and the Mediterranean (Daniel-Rops 2001). The organizational forms could range from a small household to one styled like the Pachomian type. Basilical monasteries near urban centers were also built. Legally, monasteries were brought under the control of the bishop, and monastic rules were established to reflect the development of a hierarchy of officials with specific responsibilities and duties, with the abbot, as the spiritual leader of the monastery, having increasing powers. Furthermore, by this time, monasticism was more integrated into the ecclesiastical organization whereas before it was considered a fringe group.

Monasticism increasingly became a force in the realm of education, organization, economy, and social amelioration (religion) during a time of social and political upheaval in western Europe (Cantor 1994; Daniel-Rops 2001). Key monasteries established in the fifth century A.D. in Gaul and Italy became places where bishops were drawn. By the ninth century, the basic rule (Rule for Monks) for most of all western monasteries followed the rule promulgated by Benedict of Nursia when he established his monastery in the sixth century at Monte Cassino in Italy (Lawrence 1984; Daniel-Rops 2001). The Benedictine rule continued the theme of self-sufficiency of the monastery. What this meant was that the monastery became a completely self-contained community economically, politically, and spiritually. There was supposed to be no external interference unless there were outrageous violations, and only then could the bishop or pious laypersons intervene. The abbot of the monastery had unchallenged authority to regulate the life of the monastery. Under him was a system of deans, with each being responsible for a group of ten monks. The members came from all walks of life—all classes and age groups. Monks and laymen populated the monastery. Monastic life in a Benedictine monastery was one of absolute regularity with two meals a day and care provided to the sick and aged. There were no extreme forms of ascetic practice undertaken such as that of prolonged fasting and self-flagellation. Physical labor was encouraged, and six hours were allocated for it in a monk's day. Meditation, prayers, and private reading were also part of the monk's daily life.

In the three centuries after Benedict's death in A.D. 543, the monastery underwent transformation. With the Dark Age conditions persisting in western Europe, the monastery became a key institution for medieval society (Cantor 1994; Daniel-Rops 2001). Flushed with able and literate monks it provided the necessary services to the community. In a wider context we see the decline of the state following the Germanic invasions, widespread socioeconomic distress, environmental changes, and the atomization of social and political life. The resultant emphasis then was toward local institutions. In such an environment, the monastery fitted well with this emphasis of *localism* and assumed important functions in the areas of education, economics, politics, and religion.

With all the upheavals persisting, the monastery became the repository of Christian and classical texts, and it also performed the function of producing manuscripts (Daniel-Rops 2001). Schools were also organized under the tutelage of the monasteries in view of the socioeconomic distress. With the decline of Episcopal schools through lack of church support and the disappearance of state and municipal schools, the Benedictine monastery schools provided continuity and support to education. Such endeavors meant that a large proportion of the literate population in Europe was trained at these monasteries. With its library and a steady supply of teachers, it provided an effective educational system for the early Middle Ages. In fact, during the Dark Ages, the monks were the learned men of the period.

Furthermore, the monks took on the task of preserving Christian literature and the monastery became a repository for it. By the ninth century A.D., the major monasteries had flourishing schools and libraries, and as well, a *scriptoria* for producing manuscripts. Some monasteries had added functions of not only reproducing manuscripts, they also wrote letters and diplomas on behalf of the rulers. This concentration in the letters was accompanied with the development of technical excellence in the art of calligraphy and a specialization in drawing illustrations and miniature paintings for the manuscripts (Lawrence 1984).

As the century progressed, the monasteries expanded and became financially independent and successful following the receipt of donations of manors from aristocrats for services rendered (Innes 2000). Describing the conditions in the Middle Rhine Valley, Innes (2000, 47) has suggested that the monasteries were the "'multinationals' of the ninth century." These manorial estates were highly productive, and by the tenth century, a considerable part of the most successful manorial estates in Europe was owned by the Benedictine monasteries. Their productive success was fostered by undertaking various innovations in farming practices, and the estates pioneered advances in agricultural science. With such a development, the abbots became a political force and were part of the political order of the community in which the monastery was located. Later on in the period, the monastery became a place for the employment of craftsmen and servants. It also owned houses in the local towns and villages and was the landlord in these urbanized settings. Its vast holdings of land had a fair proportion of tenant farmers as well. The latter provided rent in kind or cash. The lands of the monastery provided the food that was consumed by the incumbents of the monastery, and this included the monks, servants, and other lay persons attached to the monastery. Such continuous linkages with the local society and economy led to monasteries being less self-sustaining units, and they increasingly became more institutionalized into the social, economic, and political landscape. This was a major transformation from its origins in the Egyptian desert. The monastic membership became less diverse in terms of class groupings and increasingly drew its members from the aristocratic or higher classes. The Frankish and Anglo-Saxon aristocracies supported the monastic movement with their donation of lands as well as the entrance of their sons and daughters into the monasteries.

The great abbeys also provided a colonizing function in those areas that were newly incorporated into specific kingdoms through conquests by the provision of education and religious services. For example, they performed this function on the frontiers of the Carolingian Empire (Lawrence 1984; Daniel-Rop 2001). Politically as well, certain rulers, such as Charlemagne, used the abbots as their political emissaries while the Carolingian rulers expected the abbots to supply contingents of soldiers for their armies. This process culminated in its overall promotion by the Carolingian dynasty by the eighth and ninth centuries (Dunn 2000).

As a reaction to the turbulent period of the Dark Age of Antiquity, Christian monasticism in its early beginnings in Egypt provided an alternate worldview and life practices that were very different from the institutionalized religious order then. As time progressed and conditions changed, as an institution Christian monasticism had to adapt to its environment. It is clear by the end of the Dark Ages around the tenth century A.D., Christian monasticism as an institution was integrated into the socioeconomic, political, and religious landscape of the times. Some aspects of its original precepts remained but its original "challenge" to the then systemic order of institutionalized religious and political economic practices was sublimated as a consequence of the functions that the monasteries undertook during the Dark Age of Antiquity, such as in the areas of education, religious practices, and economic affairs. Because the monasteries had abandoned the concept of isolation and integrated themselves to the local socioeconomic environment, the interrelationship between the community and the monasteries was firmly established in this later period. On this basis, the concept of a self-sustaining community went beyond the monastery to encompass its neighborly surroundings. Functioning within a political landscape of localism, this interrelationship of the monastery, its manorial estates, and the village or town became one of the main features of the socioeconomic and political tapestry by which medieval society was structured and organized.

BIOREGIONALISM

The emergence of monasticism in the fourth century A.D. as a reaction to the turbulent conditions of the Dark Age of Antiquity reveals the development of an institution structured to deal with the systemic crisis. This response to the crisis that occurred almost 1,700 years ago appears to recur throughout the history of Western civilizations whereby alternate worldviews and practices reemerge in response to the crisis conditions of the time. In the late twentieth century, with the global crisis in diminishing natural resources, large-scale species extinction, incessant population growth, and worldwide urbanization, an alternate worldview and practice reappears in Western civilization. This movement, known as bioregionalism, started in the 1960s and 1970s and offers a worldview and life practice that is counter to the dominant paradigm that is oriented to growth and industrial progress.[4]

The early stages of the movement were based mainly around the efforts to live an alternate lifestyle that is in contradistinction to the industrial-scientific paradigm. In a similar fashion, like monasticism, bioregionalism's rise was in reaction to the socioeconomic, political, and ecological conditions of the times, though the specificities were different. Bioregionalism was a response to the turbulent times of protest in the 1960s and 1970s over the excesses of federal-state control, the war in Vietnam, the corporate control of economic life, and the overall technologization of life in general. The reaction was followed with back-to-the-land initiatives to organize life that is in direct contrast to the orthodox practices then, and to return to Nature. These initiatives provided some communities the opportunity to practice the bioregional vision (Aberley 1998; Hansson 2003).

From such origins, the movement emerged to deal with the politics of land use in relation to the locale in which the practitioners dwell be it in the rural areas or in a more urbanized setting. Coupled with the rise of other environmental movements in the 1970s and beyond, bioregionalism concretized itself through the land practices and cultural specificities of certain select communities. This vision was disseminated through various publications, gatherings, and congresses. The early proponents of bioregionalism came from different disciplines and professions. Poet Gary Snyder wrote early treatises on the idea of bioregionalism (Snyder 1974, 1980). Peter Berg, actor and writer, of the Planet Drum Foundation, was also an early proponent (Berg 1978). With Raymond Dasmann, an ecologist, Berg produced a treatise on reinhabiting California (Berg and Dasmann 1978). Other early supporters of the bioregional vision include Jim Dodge (1981), Freeman House (1998), and Kirkpatrick Sale (1991).

Bioregionalism's basic premise is that the human being is part of a living world, and as a result there *should not* be a separation between the human and other living things. This interrelationship between humans and their surroundings is considered harmonious, complex, and diverse—in short, a view of the world that is Gaean. Bioregionalism's vision in the early 1970s was a reaction to what was occurring in the twentieth century: excessive consumption of natural resources, the impending scarcity of natural resources and the overdrawing of them by humans, species extinction, uncontrollable urbanization, excessive deforestation, and the human impact on the weather. To overcome this global crisis therefore requires a radical change in our conception of our relationship to our natural surroundings and our life practices.

In reaction to national or international efforts that have precipitated crisis conditions, bioregionalism assumes a different scale of life practice. Whereas the common or orthodox assumption of scale operates at the level of state or nation-world, the bioregional scale covers region in place of state, and community in place of nation-world (Sale 1991, 50). This shift in juridical-spatial dimension assumes that for bioregionalism, like monasticism, the emphasis is on localism instead of globalism that is commonly assumed or encouraged today.

BIOREGIONAL BOUNDARIES AND LIFE PRACTICES

Bioregionalism's emphasis on localism informs life practices and focuses efforts revolving around the fundamental basis of what is termed *place* (e.g., see Aberley 1998; Carr 2004; Dodge 1981; House 1998; Sale 1991; Berg 1978; Berg and Dasmann 1978; Chew 1997; Devall 1998; Thayer 2003). For bioregionalists, place is considered to be the bioregion, with *bios* meaning life and *region* meaning the life territory (Dodge 1981). Given this, the life practice means to have responsibility to the locale where one inhabits. Such an obligation means that the essential life and natural processes of the place remain intact and evolving. By pursuing this sense of place one also develops a direct relationship with one's immediate surroundings. This relationship is then transformed into a set of life practices and awareness. It involves being aware of the rhythms and contours of the landscape and the ecological relations that underline its dynamics. It entails being aware of the diversity of life and the need either to restore what has been degraded or to ensure that ecological degradation does not take place. What it also means is the enriching of the sociocultural life and the natural aspects of the place. To this extent, early bioregionalists such as Berg and Dasmann (1978) have coined a term to depict such responsibility and life practice: *rehabitation*. Listen to how they describe it:

> Rehabitation means learning to live-in-place in an area that has been disrupted and injured through past exploitation. It involves becoming native to a place through becoming aware of the particular ecological relationships that operate within and around it. It means understanding activities and evolving social behavior that will enrich the life of that place, restore its life-supporting systems, and establish an ecologically and socially sustainable pattern of existence within it. Simply stated it involves becoming fully alive

> in and with a place. It involves applying for membership in a biotic community and ceasing to be its exploiter. (Berg and Dasmann 1978, 217–18)

To achieve this, such an awareness and responsibility require the efforts of not only an individual understanding and realization but that of the whole community. The emphasis then is on community, and it encourages the whole community to partake in such life practices over the duration of generations of community members. It becomes a life-long process for the community in question.

Given the changing scale of bioregionalism's emphasis away from the boundaries of the state-world to one of region, what then are the dimensions that depict a bioregion? Unlike the hard political boundaries depicted by state, county, or nation, the boundaries of the bioregion are considered to be soft. Dodge (1981) has proposed that the boundaries of the bioregion be demarcated by physical and biotic criteria. For him, the biotic shift encompasses moving from one region to another with changes in plant and animal life that would suggest a natural boundary. As a benchmark indicator, if about 15 to 25 percent of the animal and plant life changes from one region to another, then this would demarcate the boundaries of one bioregion with another. Other characteristics that could be used to depict different bioregions would be landforms such as watersheds and mountain ranges. In addition, since a bioregion means also a life territory, the cultural and phenomenological aspects defining a specific area would also be considered a demarcating boundary of a bioregion.

In reaction to the politics, economics, and social values depicting society in the twentieth century then and now, the bioregional vision in these spheres of life is shaped to counter what is viewed as the sources and causes of the destruction of Nature, and the strain and tension of sociocultural relations. Sale (1991) offers an articulation of the bioregional vision in these spheres governing social life.[5] In the realm of economics, the bioregional view is that the economy is based on maintaining the natural world instead of unsustainable excessive extraction of natural resources—in other words, an economy based on conservation. What this means is an economy dependent on "a *minimum* number of goods and the *minimum* amount of environmental disruption along with the *maximum* use of renewable resources and the *maximum* use of human labor and ingenuity" (Sale 1991, 69). Therefore, economic growth is not deemed to be the goal of such an economy, instead sustainability is preferred.

With conservation as an operating principle of a bioregional economy, self-sufficiency of the economy becomes the underlying principle. A self-sufficient bioregion would not only be economically stable, it would also mean that it will not have to be dependent on other regions or other parts of the world to ensure its economic reproducibility. By avoiding this, a bioregional economy would not be destructive to natural areas and social-cultural lifestyles of other regions and parts of the world as the self-sufficient bioregion would not be dependent on these resources. In this current era, such a stance is inimical to the globalizing economic tendencies of the world. With a self-sufficient bioregion based on the economic precepts listed above it would mean that all the issues, tragedies, and consequences of a growth-based economy ranging from pollution, diseases, toxicity, and crime could be avoided.

In the area of governance, decentralization as opposed to centralization is the political norm. Where diversity of all life and interdependency relationships are celebrated, the emphasis is not on hierarchical efficiency. Rather complementarity and reciprocity of relationships are fostered. It means that there is no identification of hierarchical priority of functions or rank, for all parts are necessary to the reproduction of the whole. In other words, there are complementary functions and relationships to maintain the polity as a whole. Shared responsibility becomes the norm of governance. With this form, governance, communication, and networking become necessary and important. Networks therefore ground the political governance to horizontal decision making and autonomous units of political participation.

With the celebration of diversity, it would mean that different political relations would likely exist according to their own environmental settings and their own political needs. Instead of political homogenization that is the tendency of political globalization strategies today, the stress is the autonomous evolution of political practices that are compatible to life practices and the requirements of the cultural-economic landscape.

If such is the nature of political and economic governance, the socio-cultural aspects of life would entail certain specific characteristics that embrace these thematics. The principal of the intrinsic right to evolve for all living things suggests that the diversity of culture and other life forms is encouraged and promoted. With the linkage to place, the cultural understanding of the historical relationship between the human community and the land is understood and remembered. This implies that past indigenous cultural experiences are known, and if possible, practiced. This cultural foundation is built upon by the present indigenous configurations with the aim of considering future development as "future" primitive (Aberley 1998). Cultural practices and development thus are evolving and adapting to the biological integrities of the landscape.

BIOREGIONAL OUTCOMES

Various indicators of bioregionalism are evident since its emergence. Literature written by bioregionalists and on bioregionalism by specialists has been published (e.g., Aberley 1998). They have focused either on the philosophy and perspectives of bioregionalism or on specific case studies of bioregional projects (e.g., House 1998). In terms of quantity of publications beyond scholarly books, there are over 60 bioregional publications including another 140 publications that have materials on bioregionalism (Carr 2004). Information dissemination by no means is the only outcome of this bioregional vision. As bioregionalism is a perspective that includes a practice that has a sense of life experience, practice outcomes undertaken by bioregionalists have included ecological restoration projects covering forests, streams, watersheds, and species. Besides restoration, bioregionalists have undertaken mapping assignments, education, planning, land use, and renewable energy projects (Carr 2004). Organizational activities include promoting eco-living, barter-exchange networks, community currency, land trusts, and volunteer services. Developing a sensitivity and appreciation of the cultural configurations and experiences of a bioregion, and promoting the cultural more-ways are also other outcomes of the bioregional vision.

Life practices undertaken by different bioregional communities with a bioregional vision are shared through networks and congresses. For example, the Ozark Area Community Congress has been meeting each year since 1980. These regional meetings, such as those held in Kansas and the Pacific Northwest, have led to convening of the North American Bioregional Congress with the first meeting held in 1984. There have been eight such national congresses attracting participants from other parts of the world (Aberley 1998; Carr 2004). Such meetings and interest shown in the bioregional vision are visible in the bioregional groups that have formed. According to Sale (1991) by the early 1990s there were more than sixty bioregional groups established. The bioregional vision has also spread to other environmental groups that have been formed to protect the environment. The principles of bioregional living and the sense of place and home have been embraced by other environmental groups and associations (Carr 2004).

The extent of the impact of bioregionalism can be seen in the sphere of public life. Bioregional ideas have been included in planning activities of cities and state agencies in North America as a consequence of bioregionalists' work in attempting to rehabitate their life places. For example, the California Resources Agency has divided the state of California into bioregions for planning purposes and in order to manage the natural resources of the state of California.[6] Federal and state agencies responsible for natural resources and wildlife have also included the concept of bioregion and watershed distinction in their planning activities. In urban settings, bioregional groups have conducted activities to have metropolitan centers such as San Francisco, Vancouver, and Toronto plan and undertake civic projects with a bioregional vision in mind.

TWO ALTERNATE LIFE PRACTICES DURING CRISIS PERIODS

Monasticism and bioregionalism as outcomes to system crisis appeared during periods when there were tremendous socioeconomic, political, and ecological upheavals. Both provided alternate worldviews and life practices with an emphasis of returning to past interpretations of sociocultural life and Nature. These movements were not proponents of growth, consumption excesses, and centralization, for they were reactions to the tendencies that seemed to underline human societies for the last five thousand years of human history and that reached certain crisis points during Dark Ages. Instead, their emphases were on localism and self-sufficiency, which entail operating with a different set of operating parameters. Instead of kingdom or empire, the community was promoted. This notion of community was extended further as monasticism and bioregionalism developed to encompass a wider notion of community as a consequence of historical changes. In the case of monasticism, it extended beyond the monastery to include the local villages and towns that were in the vicinity of the monastic order when manors and lands were donated to the monastery as the movement became further integrated into the political and economic order. Bioregionalism too had to coexist with its immediate surroundings populated by others who did not share the bioregional vision. Over time, through life practices, protests, and communication, notwithstanding the ecological crisis conditions that reinforce the bioregional concerns, some aspects of the bioregional vision has been incorporated into the public sphere of planning and discourse.

Christian monasticism continues today. It appears during a period of crisis, and we have witnessed its evolution through world history. We have noted its outcomes, and the impact it had during those turbulent times during the Dark Age of Antiquity. As a movement, it offered in its early formative years a reorientation of human conception of the world and consequently a practice followed. During times of political and socioeconomic disorder, it continued the traditions of the distressed communities and was a repository of knowledge. As an institutional structure over time, it was eventually absorbed into the ruling political and socioeconomic order, and as a result it has continued to this day though not in the same distinction and rationale as it had when it emerged during the fourth century A.D.

Bioregionalism's historical development has yet to be completed. Like monasticism, it also emerged during a period of political, socioeconomic, and ecological crisis. Offering a worldview and life practice that is different from what existed (and exists today), bioregionalism as a life practice has taken root only in isolated rural communities and in certain pockets of urban settings. The bioregional vision of localism, self-sufficiency, place, and home increasingly resonates in an era of globalism that promotes exactly the opposite of what bioregionalism proposes. At this *conjoncture* of world history, the anticipation is that as the effects of globalism come to be felt and revealed as history unfolds; the bioregional vision will perhaps offer a viable alternative to the globalization-growth model that is currently in play. Of course, there are many other alternatives as well in terms of options (e.g., The Millennium Project, UNCED Report, etc.). Furthermore, with the current emphasis on consensus building as an approach toward governance, especially in the arena of natural resource use and planning, the possibility of cooptation of bioregional ideas and practices into mainstream environmental decision-making processes and procedures is very likely. Just like monasticism after its establishment and growth, there is no reason to believe this will not happen.

What choices and practices we undertake will depend on the structural conditions of the planet as this millennium unfolds. We know that Christian monasticism emerged during the Dark Age of Antiquity. From our past history, we must also realize that human choices are often made due to crisis conditions and that we often choose because of the contingencies we encounter. Is this the same fate that we will face in the future?

NOTES

1. Dunn (2000, 2), however, has stated that the land was very fertile.
2. These developments were discussed in detail in *The Recurring Dark Ages*.
3. The eremitical way of life adopted by desert hermits took its name from the Greek word *eremos*, which means desert.
4. There are other similar types of movements in other parts of the world, such as in India (e.g., Bahuguna 1996).
5. See also Carr (2004) in terms of an account of bioregional governance and organization.
6. Unfortunately, the boundaries of the identified bioregions, though following the principles of the boundaries of a bioregion, end at the official political boundaries of the state of California.

3

The Transitions

System transition occurs when the reproductive capacity of the system has reached its limits. In terms of historical social systems, such occurrences are rare. In an era of crisis, the puzzling question often posed these days is whether our historical social system is at the point of transition.[1] A deliberation on system transition is important because it encourages us to analyze, understand, and project the likely socioeconomic, political, and ecological configurations that our human communities will possibly face in the future. To do this, in my view, will require an assessment of the past, especially during periods of system transition, so that we can project and assess what historical alternatives are possible, and the constraints that are likely to appear as well.

To what past do I use as a point of reference so that I can anticipate our possible futures? As I have stated in previous chapters and in *The Recurring Dark Ages*, Dark Ages are significant periods because they are *conjonctures* of system transformation. They are historical moments that set into motion a phase of reconfiguration and adaptation that structures the system as a whole in the long run. Dark Ages are critical crisis periods in world history when environmental conditions play a significant role in determining how societies, kingdoms, empires, and civilizations are reorganized (Chew 2007). They are also periods of devolution of human communities, and as such, from the perspective of human progress, a period of socioeconomic and political decline, decay, and retrogression. This devolution means a slowdown of human activities resulting in a period of rejuvenation and recuperation for Nature instead of the uncontrollable utilization of natural resources that normally occurs when there is expansion. In short, Dark Ages are periods of restoration of the natural landscape. Therefore, two processes come into play during these periods of darkness: the rejuvenation of Nature and the reconfiguration of socioeconomic and political systems.

The process of the rejuvenation of Nature means the restoration of the natural landscape and, over the duration of Dark Ages, makes available the necessary natural resources and energy for recharging the reproductive capacity of the social system. At the social-system level, human communities impacted by these Dark Ages are forced by circumstances, and not perhaps by choice, to reorganize in a different manner to meet the contingencies of ecological scarcity, diseases, and climate changes. These reorganizations and reconfigurations over this period of the Dark Ages, if we examine the past history of historical events and circumstances, are also historical phases of innovations, learning, and discoveries. The convolution of these historical processes and actions forms the basis for system transformation and the evolution of the world system.

SYSTEM CRISIS: TRENDS AND TENDENCIES

Crisis emerges when the social (world) system encounters a scarcity of natural resources that it requires in its system maintenance and reproduction. By no means is this the only threat. Changing weather, natural disturbances, and diseases can also be considered as posing barriers to the overall reproduction of the world system. All these elements can be placed in the natural system column of conditions that impact the continuance of the world system. Socioeconomic and political conditions belong to the other column under social processes that are also important components necessary to world system reproduction. The interactions between the social and natural generate the energy, resources, structures, and processes that power the reproduction of the world system. System transition and transformation occur when limits are reached in the social and natural systems and/or breakdown between natural and social systems relations. Such limits can include natural resource scarcity, changing ecological and natural conditions, socioeconomic decay of institutions and processes, wars, migrations, and loss of political legitimacy.

NATURAL SYSTEM LIMITS: FROM THE IMMEDIATE PAST TO THE PRESENT

For this current era, the early warnings that the natural system has limits in terms of its provision of resources for the reproduction of the social system appeared in the work of the Club of Rome in the early 1970s (Meadows et al. 1972; Mesarovic and Pestel 1974). The principle announcement of the Club of Rome then was that if the growth trends in "world population, industrialization, pollution, food production, and resource depletion continue unchanged, the limits to growth on this planet will be reached sometime within the next one hundred years. The most probable result will be a rather sudden and uncontrollable decline in both population and industrial capacity" (Meadows et al. 1972, 29). Exponential growth in population and the economy according to the Club of Rome report are the key factors in accelerating the consumption of the available natural resources and, in the process, widening the absolute gap between rich and poor nations. Uncontrollable consumption as a consequence of exponential population growth and the widening disparity between rich and poor nations provide the conditions of instability for the social system. At that time, the rate of natural resource usage was growing faster than the population, and it still is the case now. What this suggests is that the population growth and the increase in capital stock were driving the exponential curve of resource consumption. The end result, according to the Club of Rome then, was that there were limits to the exponential growth that the human community was displaying and ignoring. Furthermore, the outcome of the uncontrollable economic growth was global pollution. According to the Club of Rome, some of the measured pollution levels, such as carbon dioxide and toxic wastes, were growing exponentially and that in the long run, through a feedback loop, would also pose limits to growth. Given such trends, the Club of Rome concluded that economic growth will reach its limits well before year 2100.

In the 1974 follow-up report, *Mankind at the Turning Point: The Second Report to the Club of Rome*, the global warning announced in the first report had turned into an alert of the various crises that the world community faced. The crises covered the areas of population, environment, food, energy, and raw materials. This is how it was framed:

> New crises appear while the old ones linger on with the effects spreading to every corner of the earth until they appear in point of fact as global, worldwide crises. Attempts at solving any one of these in isolation has proven to be temporary and at the expense of others; to ease the shortage of energy or raw materials by measures that worsen the condition of the environment means, actually, to solve nothing at all. Real solutions are apparently interdependent; collectively, the whole multitude of crises appears to constitute a single global-crisis syndrome of world development. (Mesarovic and Pestel, 1974, 1–2)

Given this scenario, the second report suggested that the world community can only resolve these crises at the global level, the solutions cannot be achieved by traditional means, and that the crises were not temporary but long-term.

This alert on natural resource scarcity and the crisis of the environment was followed up more than a decade later in the 1987 report of the World Commission on Environment and Development titled, *Our Common Future*. Commonly known as the Brundtland Report after the Commission's chairperson, Gro Harlem Brundtland, the report echoed the alert warnings sounded earlier by the Club of Rome in terms of what prospects were facing the global human community at the end of the late twentieth century. Building on the United Nations Charter for Nature proclaimed in 1982, the usual signals of environmental degradation—natural resource scarcity, species extinction, atmospheric pollution leading to global warming, deforestation—were identified and empirically presented. Sharing the same view as the Club of Rome, the Brundtland Report underlined some of the factors that foster environmental problems such as world poverty and international inequality. Poverty is targeted as "a major cause and effect of global environmental problems" (World Commission on Environment and Development 1987, 3). As with the Club of Rome, the commission outlined the character and extent of the condition of species and ecosystems. Even in 1987 there was already scientific consensus on the rate of species extinction never before witnessed in the history of the planet. Habitat alteration was being conducted on such a global scale that it posed a grievous threat to biodiversity. Besides an impending biodiversity crisis, the report warned of the scale and pace of deforestation and noted extinction patterns that were unprecedented on a global scale in world history. Associated with this was the likelihood of climatic changes, and along with the accumulation of greenhouse gases, global warming was anticipated in the early parts of the twenty-first century. The global warning of the Brundtland Commission also alerted us to the impending crisis of global energy resources.

Aware of the fact that the primary energy resources we use are nonrenewable, the commission warned us of the finite nature and the impending scarcity that we face. The scale of global threat is further increased in the area of global warming and environmental risks with the nonrenewable energy sources that we use. Of these threats, the Brundtland Commission identified four that have severe repercussions for the environment: (1) carbon dioxide emissions from the burning of fossil fuels leading to climate change; (2) urban-industrial air pollution caused by atmospheric pollutants; (3) acidification of the environment from pollutants; and (4) the risk of nuclear reactor accidents and the problems of waste disposal and the dismantling of reactors after their service life is over (World Commission on Environment and Development 1987, 172).

The warnings of global natural-resource limits were articulated most forcefully via a global conference organized by the United Nations. Five years after the Brundtland Commission reported its findings, the United Nations Conference on Environment and Development was held in Rio de Janeiro in 1992. Commonly referred to as the Earth Summit, this global meeting identified the natural resource limits and scarcity the global community faced and the causes and consequences of such tendencies. The identified natural resource scarcities and outcomes, the pollution levels, and consequences as a result of uncontrolled human economic and population growth announced and discussed at the conference were by no means new insights to what the global community was already aware of or choose not to hear. Building on the warnings of the Club of Rome reports and the Brundtland Commission, the Rio conference simply reiterated the dangers and sought for a global agreement to handle these natural resource and pollution crises that the global community and world economy faced. With the impending shrinking availability of natural resources and global warming that will impact socioeconomic life, the solution was a call for *sustainable development* as the theme that all participating nation-states at the conference should adopt in their national programs and policies. One of the outcomes is the Rio Declaration on Environment and Development that contained twenty-one principles for action—Agenda 21.

This impending natural resource scarcity issue including deforestation of the planet was repeated again—especially the latter act of degradation of the environment—by the World Commission on Forest and Sustainable Development in 1999. As with all reports of world commissions, alarms and alerts were raised buttressed with data on the rate of degradation and the impending disasters ahead. According to the commission, deforestation had reached extreme levels as a consequence of the previous phase of incessant growth of the world system. The ferocity of deforestation globally has confirmed the pace of cutting during the twentieth century. Soil erosion, flooding, and species endangerment that are often the outcome of deforestation were also signaled. Another outcome of global deforestation is the impact on global warming and this was also underscored by the commission. As a summary, the deforestation crisis had reached unacceptable limits according to the commission. At the end of the twentieth century, forests had virtually disappeared in twenty-five countries, eighteen have lost more than 95 percent of their forests, and another eleven have lost 90 percent.

This global issue of natural resource scarcity has been followed up further with updating by the original team that projected limits to growth to the Club of Rome over three decades ago. In two separate published volumes, *The Limits to Growth* and *Limits to Growth: The 30-Year Update*, Donella

Meadows and her colleagues again reiterated the warnings of natural resource scarcity and pollution issues that are degrading the landscape and the threats they pose to the future survivability of the global human community.

In these two sobering volumes, Meadows et al. (1972, 2004) calculated that we have overshot the limits of our planet's natural assets in terms of sustainability, and collapse could follow in the twenty-first century if there is no change in human practices. Such crisis conditions and levels as "the decline in oil production within important nations, the thinning of the stratospheric ozone, the mounting global temperature, the widespread persistence of hunger, the escalating debate over the location of disposal sites for toxic wastes, falling groundwater levels, disappearing species, and receding forests," to name a few, are the critical threats to the reproduction of the social system (Meadows et al. 2004, xvii).

The most recent assessment coordinated at the global level (Millennium Ecosystem Assessment) by the United Nations in response to numerous governments' request for information continues to report such scarcities of natural resources. According to the Millennium Ecosystem Assessment (2003, 2005), threats to the ecosystem continue. For example, updating the assessment of the World Commission on Forests and Sustainable Development (1999) on the state of the world's forests, the Millennium Assessment has estimated that 40 percent of the world's forest cover has been reduced in the last three centuries with three-quarters of it occurring in the last two. Twenty-five countries have lost their forests completely with another twenty-nine losing more than 90 percent of their forest cover. In the tropics, 10 million ha of forest cover are lost annually. This order of forest removal is accompanied by loss of species in the higher taxa amounting to between 12 and 52 percent. It is estimated that 24 percent of the 4,700 mammal species, 30 percent of the 25,000 fish species, and 12 percent of the 10,000 bird species are in danger of extinction (Meadows et al. 2004). For plants, 34,000 of the 270,000 plant species are at risk. Overall, the average species population has declined by more than one-third since 1970.

It is clear from all these global assessments by world commissions and global institutions that reductions of requisite natural resources are occurring. These trends and tendencies mean that the natural resources required for the reproduction of the world (social) system is getting more and more limited.

In addition to the reports of world commissions and global institutions, other notable independent scholars have also rung the bells of crisis. Harvard biologist, E. O. Wilson's *The Future of Life* (2002), approaches the ecological degradation and crisis from the view of the natural sciences. Indicting humanity, Wilson joins the many in noting how we have decimated the natural environment and drawn down the nonrenewable resources of the planet. In doing so, according to Wilson, we have accelerated the destruction of ecosystems and caused the extinction of species, some that have been here for at least a million years. Along a similar vein, but more comprehensive in nature, is Paul and Anne Ehrlich's *One With Nineveh*. The Ehrlichs' 2004 book mirrors the Brundtland Commission's report and Earth Summit '92 except that the Ehrlichs, writing almost twenty years later, identified environmental trends and tendencies that have changed somewhat to a situation that is even more critical. With the exception of population trends, what was discussed in the Brundtland Commission and Earth Summit '92 was repeated albeit with different emphases reflecting the interests of the Ehrlichs. Also published in 2004 James Speth's *Red Sky at Morning* repeats the trends of all the previous studies. As such it confirms the scenario of a natural environment devastated ecologically through deforestation, pollution, landscape transformations, and species extinction. The condition is made worse with climate changes and an approaching natural resource scarcity, especially in the area of nonrenewable energy.

The latest summing up of this series of independent scholarly assessments of the state of the world's natural resources and ecosystems is Lester Brown's *Plan B 2.0* (2006). Brown's alerts and warnings are even more alarming. Besides identifying the usual array of impending system degradations and scarcities in the areas of petroleum, forest, water, and landscape, his singling out of the People's Republic of China (PRC), and much less so India, in terms of excessive global consumption in basic materials and food in the reproduction of the Chinese social system, thus overtaking the United States except in the areas of oil and corn, is troubling to us from the ecological perspective rather than a viewpoint based on equity. This means the level of depletion of natural resources at the global level is accelerated, especially in light of China's population, which is projected to be about 1.45 billion persons by 2031. Listen to Brown's (2006, x) worries about PRC's projected consumption level and its impact on the planet:

> What if China catches up to the United States in consumption per person? If China's economy continues to expand at 8 percent per year, its income per person will reach the current U.S. level in 2031. If we assume that Chinese consumption levels per person in 2031 are the same as those in the United States today, then the country's projected population of 1.45 billion would consume an amount of grain equal to two-thirds of the current world grain harvest, its paper consumption would be double current world production, and it would use 99 million barrels of oil per day—well above current world production of 84 million barrels.

The forecast of impending scarcities in fossil fuels, water, and rangelands engendering systemic difficulties for social system reproduction cannot be more somber (Brown 2006). The growing global water deficit as a result of demand tripling over the last fifty years has led to drained aquifers. The deficit in the long run will impact crop production. Whether they are replenishable or nonreplenishable aquifers, they are being drained to meet the increased demand for water for urban use and agricultural irrigation. In some places like Saudi Arabia, the draining of the aquifer will mean the end of agriculture. The deep aquifer under the North China Plain has been dropping about three meters per year due to excessive pumping. As a result in some places of the plain, wheat farmers are pumping for water to a depth of three hundred meters. With reduced level of water availability and impending dangers of global warming, such coalescence of conditions would ultimately lead to reduced harvest and hence social system reproduction issues. According to Brown (2006), the water shortage has already led to reduced yield for China in wheat and rice production. Harvest deficits have been addressed to date by grain imports from areas where there is more abundant irrigation water for the growing of crops. On a global basis, this means that the issue of water deficits is being transferred from areas less impacted by water shortages for irrigation to areas that suffer water deficits, such as the Middle East, North Africa, China, India, Pakistan, and Mexico (Brown 2006).

Besides drained aquifers, other water sources such as rivers and lakes have also suffered shrinkage. Some of the world's major rivers such as the Colorado in the United States, the Yellow River in China, the Nile in Egypt, the Indus in Pakistan, and the Ganges in India have experienced reduced water flow and in some years have failed to reach the oceans or seas. For example, the Yellow River has often not reached the Bohai Sea since 1985. The Nile in Egypt, which has been for millennia the life line of Egyptian farmers, now barely makes it to the Mediterranean in some years. Lakes have also been disappearing, which is quite understandable. With reduced river flows and increased pumping of the aquifers, naturally lakes will feel the impact. Lake Chad in Africa, the Aral Sea, and the Sea of Galilee have all shrunk. In terms of national level disappearance of lakes, the Chinese experience is quite alarming. In western China's Qinhai province, over the last twenty years about 2,000 lakes have disappeared from a total of 4,077 (Brown 2006).

The aforementioned pages outlining the conditions of natural resource scarcities and availability for system reproduction as outlined in the various studies and reports by Club of Rome, world commissions, and so forth leading to a global environmental crisis have been questioned by some such as Bjørn Lomborg in his 2001 book, *The Skeptical Environmentalist: Measuring the Real State of the World*. Unfortunately, Lomborg, in this statistically leaden volume, tries to disprove the issue of the scarcity of natural resources and so forth instead of scrutinizing the studies published by the Club of Rome, the Brundtland Report, the Meadows et al. Limits to Growth reports, the World Commission on Forest and Sustainable Development, and Intergovernmental Panel on Climate Change, chose to "zero in" on a nongovernmental environmental organization: the work of the Worldwatch Institute (WWI) and the writings of its director, Lester Brown. Indicting the WWI and Brown for their value-leaden projections and prognoses on the state of the world, Lomborg (2001, 32) falls into it himself: "My claim is that things are *improving* and this is necessarily a discussion which has to be based on facts" (emphasis in original). With this claim, he proceeds to marshal, according to him, the available "facts." The problem is with him presenting his "facts" and the way he interprets his facts within the value-leaden "neo-environmental" position he has taken. Lester Brown who is an environmentalist wants to signal the global environmental dangers we face. Lomborg, on the other hand, though he does not admit of it, has his own agenda and values. The first is a positivistic belief in that statistics and science can be the final arbiter of any debate, and secondly, the world is improving instead of deteriorating.[2] This agenda is clearly reflected in his chapter on the state of the world's forests.[3] In this case, he concluded that the world's forests are not under threat, and he indicted the Worldwatch Institute for stating so. Unfortunately, Lomborg does not refer to or challenge the report of the World Commission on Forest and Sustainable Development, which also stated that the world's forests are disappearing.[4] Furthermore, recently the Millennium Assessment on the state of the world's forestry has also underscored the dangers that the World Commission on Forest and Sustainable Development announced over five years ago for the end of the millennium.

Despite Lomborg and his "neo-environmental" philosophy, the various recent reports such as the Millennium Assessment and the World Commission on Forests and Sustainable Development clearly point to systemic natural resource scarcities in various spheres necessary for the reproduction of the social system, notwithstanding the natural system's own ability to regenerate. How does this bode for the future reproduction of the world system especially when Nature-Culture relations are likely to be disrupted continuously as the current millennium proceeds?

SOCIAL SYSTEM CHANGES

Wars and Political Upheavals

If history has anything to tell us, times of scarcity in either the natural or social realm generates upheavals and anxiety about the future. For example, if we examine the conditions during the Dark Age of Antiquity over 1,600 years ago we find such exhibition of strains of social disequilibria. As the previous volume, *The Recurring Dark Ages*, has revealed and demarcated: there were political upheavals in Asia, and in both the western and eastern portions of the Roman Empire, deurbanization in the Roman provinces, slowdown in population growth in Europe and Asia, heightened taxation in the Roman Empire before its collapse, migrations within and without the Roman Empire, wars, short reigns of Roman emperors and political struggles in China, and the increasing influence of religion in Europe in every aspect of social life, including education and the sciences, in the latter periods following the collapse of the Roman Empire. These unsettled socioeconomic and political conditions combined with changing weather patterns, diseases, land degradation, and soil erosion turned into an implosive deterioration of the social system in place.

If political upheavals and wars punctuated the Dark Age of Antiquity, is there a pattern we can distinguish whereby during periods of Dark Ages we find an increasing frequency of wars? Unfortunately, time-series data available to examine this is sketchy and limited especially for the early periods (Eckhardt 1992). With a limited time series and data coverage for the early period of world history from 2000 B.C. to A.D. 1 (see figures 3.1–3.5), we can distinguish some *tendencies* in terms of the frequencies of conflicts or wars in relation to periods of Dark Ages, though it seems that conflict conditions also occurred during periods of non–Dark

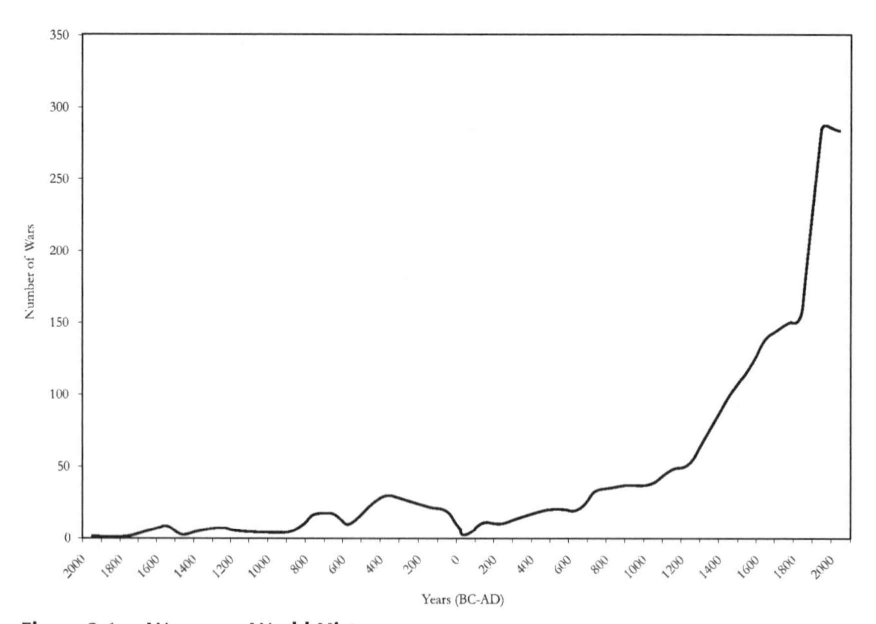

Figure 3.1. Wars over World History

Ages. Figure 3.1 covers the frequency of wars over world history. It seems that between 1000 B.C. and 700 B.C. and again between A.D. 400 and A.D. 900 there was an increased number of wars. Both of these were periods of Dark Ages. Beside this increase in conflicts during these two previous Dark Ages, it seems that warfare was on the rise throughout the world from A.D. 900 onward and grew increasingly.

In figure 3.2 covering the frequency of wars in Europe (2000 B.C.–A.D. 2000), for the period between 500 B.C. and A.D. 100, the frequency of wars declined. This period of fewer wars and conflicts also occurred after the final collapse of the Bronze Age, which was characterized by socioeconomic expansion and political efflorescence. However, frequency of wars increased from A.D 100 to A.D. 500 only to decline again until A.D. 800. Correlating this period of A.D. 100–A.D. 500 suggests that this sequencing covered the start of the Dark Age of Antiquity and a lengthy time in which Dark Age conditions pervaded Europe. Again, the frequency of conflicts increased after A.D. 800, with a respite after A.D. 900, only to start increasing again after A.D. 1000 and decreasing again from A.D. 1200. The increase of conflicts was considerable from A.D. 1300 onward, only to decrease from A.D. 1600 till A.D. 1800. Following this period, the increase of conflicts started to increase again. What about the other parts of the world?

Figures 3.3–3.5 outline time series of war frequency (2000 B.C.–A.D. 2000) for the Middle East, Far East, and South Asia. The distribution of wars in

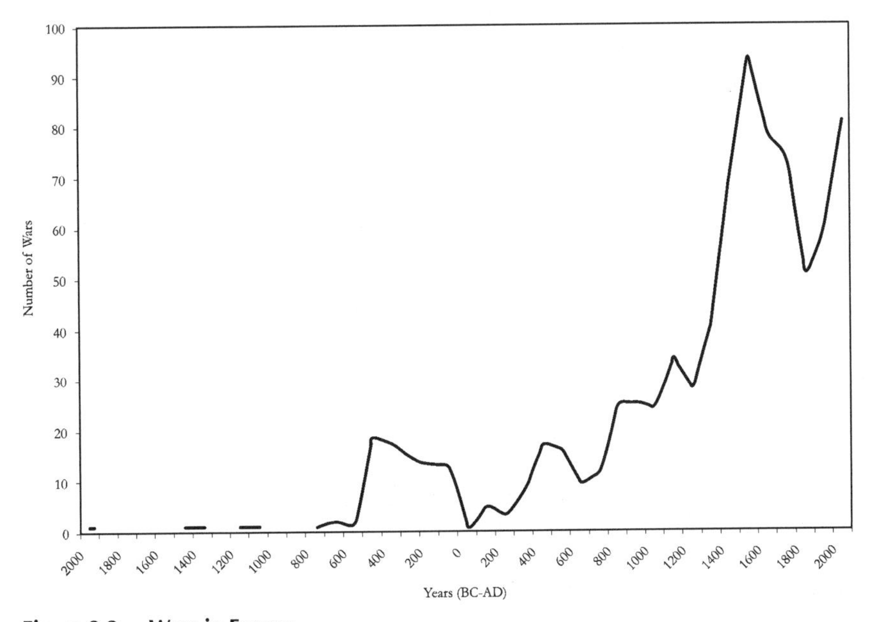

Figure 3.2. Wars in Europe

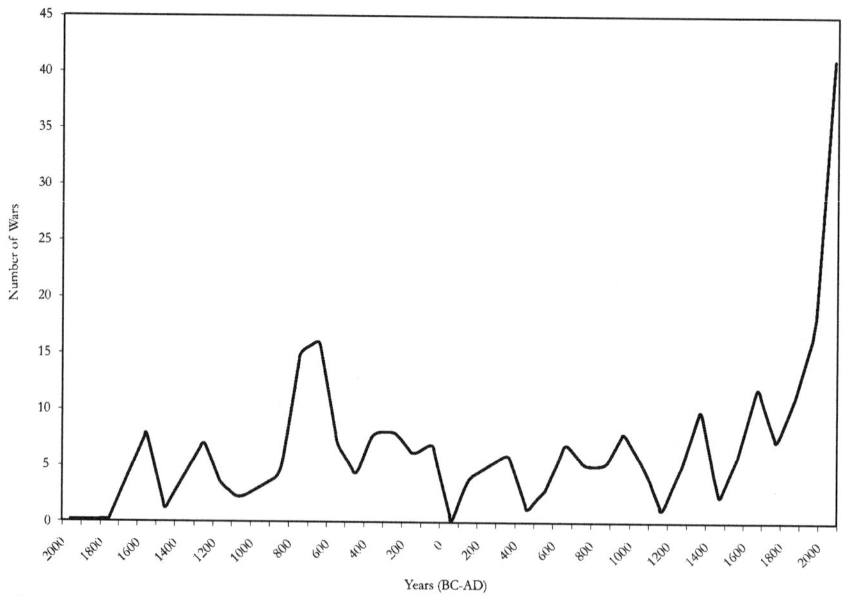

Figure 3.3. Wars in the Middle East

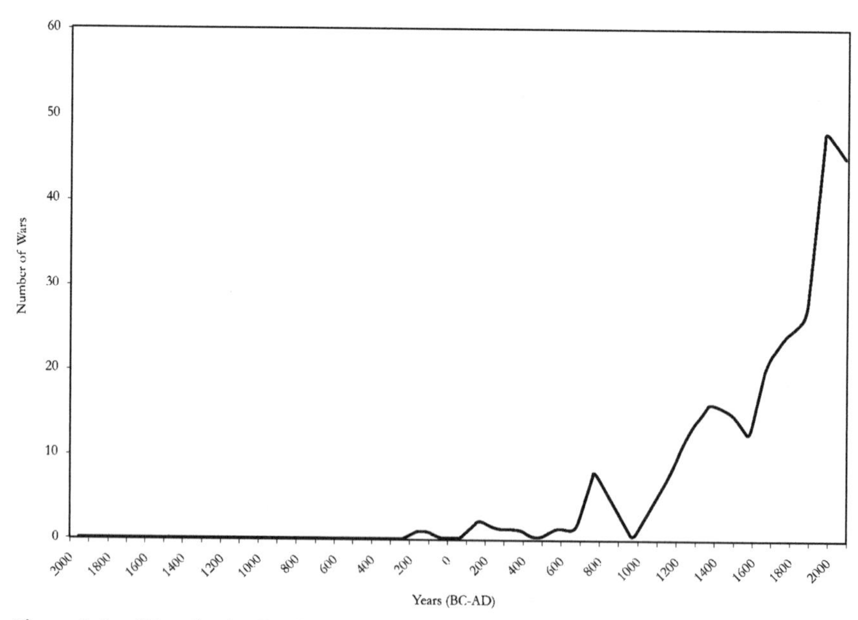

Figure 3.4. Wars in the Far East

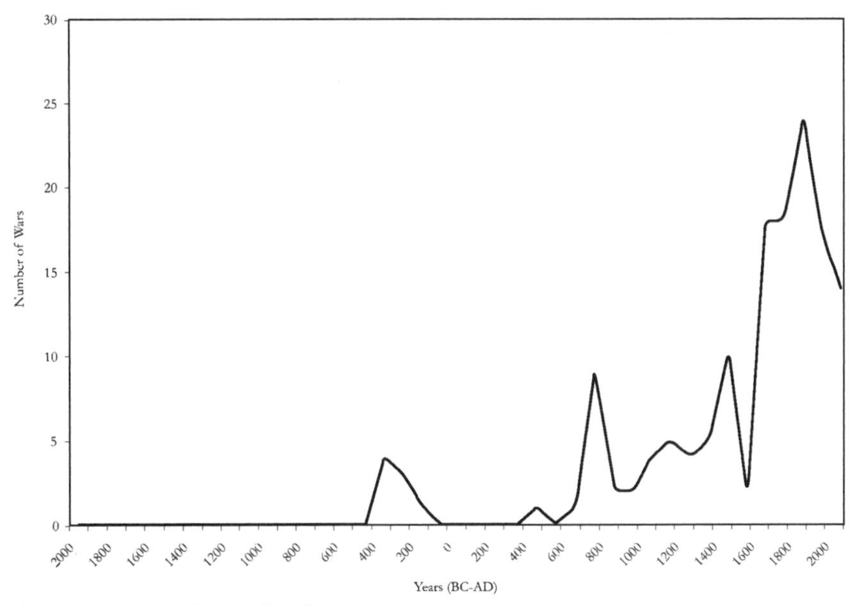

Figure 3.5. Wars in South Asia

Middle Eastern conflicts (see figure 3.3) shows an increase in wars from 800 B.C.–600 B.C. during the final collapse of the Bronze Dark Ages (1200 B.C.–700 B.C.) with the prior period from 1300 B.C. onward showing a reduction in wars. Unlike the reduction in the number of wars in Europe following the end of the Bronze Age when warlike activities subside, in the Middle East, wars emerged again from 400 B.C. onward until first century A.D. with the frequency receding until the end of the century. We see the increase in the number of wars from A.D. 100 onward until A.D. 700 and another phase A.D. 850–1000 that dovetails with the Dark Age of Antiquity, with a decrease occurring only during the fifth century A.D. It seems that beyond increases during Dark Age periods, for the Middle East after the Dark Age of Antiquity, there were also other periods of conflict that do not fall during Dark Ages. There were increases in warfare from A.D. 1200 to 1350 and from A.D. 1500 to 1700.

For the Far East (see figure 3.4), the frequency of wars increased between A.D. 700 and A.D. 900, with the prior periods showing very few wars. Conflicts started to accelerate again from A.D. 1000 until A.D. 1400. Following this was a decrease in warlike conditions until A.D. 1600. From then onward again conflicts were on the rise. The trend in South Asia (see figure 3.5) is different from the Far East. We distinguish an increase in the number of wars between 400 B.C. and 300 B.C. followed with a decline until A.D. 400. Wars were on the increase in South Asia from A.D. 600 to A.D. 800, which dovetails with the Dark Age of Antiquity. Conflicts emerge again around A.D. 1000 and continue to increase until A.D. 1500 with a slowdown between A.D. 1200 and 1300. Following this period, an increase in conflict started from A.D. 1600 to the modern era.

With limited documentation tabulating the frequency of wars, it is clear from my analysis of this available data that the historical tendencies of the occurrence of increased war frequency during long periods of economic and ecological crises (most likely a consequence of a fight over decreasing resources and accompanying climatological changes) is not as determinative as one would like to see. Despite this, what is revealing is that wars do increase during Dark Age periods as we have identified previously. Therefore, even though conflicts do occur during non–Dark Age periods, there is no doubt as well that we can expect increased levels of political conflicts or wars especially prior to and during periods of future Dark Ages.

What about our present conditions? Are there signs of economic and political upheavals that could signal the beginning of Dark Age conditions similar to those we observed during previous Dark Ages? If one takes the twentieth century as the starting point, the period has been one fraught with socioeconomic and political turmoil.[5] The century began with a world war (World War I) and ended with a regional war (Persian Gulf War). In between, there was another global war (World War II) and a number of regional conflicts (Russo-Japanese War, Italo-Turkish War, Korean War, Indo-Pakistani War, Vietnam War, Falklands-Malvinas War, Iran-Iraq War, Arab-Israeli wars, Persian Gulf War). It was also the century of changing hegemons from Great Britain to the United States, including the possible decline of the latter toward the end of the century, and perhaps changing hegemonic regions from the European West to the Asian East (e.g., see Kennedy 1988; Ferguson 2004, 2006; Frank 1998). Other social and political disruptions include the collapse of the socialist-communist experiments in the former Soviet Union and the People's Republic of China. Such political turmoil and wars usually occur during periods when hegemonic powers in the world system are on the decline and when social-system limits are in need of legitimization and reconfiguration (see Amin et al. 1982; Amin 1992; Frank 1981; Habermas 1974).

As a comparison, during the previous Dark Age of Antiquity, Rome from the fourth century A.D. onward had to experience such challenges from its neighbors, such as the Persians, Goths, Huns, Alans, and Franks, that bordered its empire (Heather 2006). The war campaigns of these groups were between Rome and them (singularly or combined) and among themselves at times. Rome had to make other social system adjustments to accommodate the social invasions from some of the "barbarian" tribes of Europe and Central Asia. Under pressure from these invasions, treaties had to be signed between Rome and these tribal groups to ensure the territories of Rome were recognized, and in times of war with other powers, Rome could rely on these barbarian groups for their support in the form of troops. Treaties were sealed with the provision of yearly endowments of gold to these barbarian tribal groups.[6] At times Rome reneged on this, following which conflict arose. In addition, the employment and inclusion of these groups within the Roman military was also another avenue whereby the Roman Empire addressed its social recruitment crisis for its legions. From the third century A.D. onward, such initiatives to placate the barbarians and to maintain Rome's increasingly declining hegemonic power in Europe, North Africa, and the Ancient Near East further generated other destabilizing conditions in the area of fiscal stability and taxation schemes. Inflationary pressures forced the dilution of the money supply that led also to other economic consequences (Jones 1959, 1964, 1974).

What were the factors that conditioned these political upheavals and wars during the Roman period of the Dark Age of Antiquity? Most often, factors such as the quest for power or glory, the political-economic need to expand territorial boundaries, or political and economic contacts leading to rivalry and competition are deemed as the most common factors precipitating wars and conflicts. For example, the recent thesis of Heather (2006) for the fall of the Roman Empire and the number of wars from the fourth century A.D. onward rests primarily on political-economic rivalry of barbarian tribes and their relationships with the Roman Empire. There is little consideration given by Heather (2006) to the decline of natural resources, climate changes, and the relationship between the natural and the social systems in trying to account for the increase of wars and conflicts between Rome and the barbarians, and the latter, between themselves. As we have discussed extensively in *The Recurring Dark Ages*, there were signs of the deterioration of the landscape as a consequence of the Roman abuse of the environment, climate changes forcing the migration of nomadic tribes from Central Asia and within Europe, and lowered revenues from taxation as a consequence of lower harvests as a result of climate changes and soil degradation. All these conditions led to social-system stress and reproduction difficulties for the Roman Empire. Notwithstanding this, there is no doubt that political, economic, and social rivalries as discussed by Heather (2006) did also play a part in the final collapse.

Such circumscribing of the conditions and factors leading to wars and upheavals direct us to consider contemporary circumstances and conditions and whether they are similar in nature, thereby conducing conditions for or of another Dark Age. During the last century, political upheavals and wars have occurred as we have listed briefly in previous pages. It is clear that some of the major conflicts were over natural resources such as oil. In the case of wars and regional conflicts fought during the twentieth century in the Middle East (e.g., the Persian Gulf War and Iraq War), Japan's entry into World War II, and Germany's annexation of Austria and Czechoslovakia prior to World War II, one can conclude that there is a strong basis that these conflicts were over access and control of natural resources. In Japan's case, its dependence for imported raw materials heightened as its industries recovered in the 1930s. Large quantities of iron ore were required to meet its steel manufacturing industry with the country's economy increasingly geared for military production. Besides coal and copper, crude oil was also in high demand as an import (see Kennedy 1988). Naturally, China and Southeast Asia were Japan's sought-after resource bases.

In the case of Germany to meet its militaristic economic strategy, natural resources and precious metals (gold and silver) were needed along with foreign-exchange reserves, after it had started to draw down its reserves in the 1930s. The acquisition of Austria gave Germany not only $200 million in gold and foreign-exchange reserves but also iron ore and oil fields (Carroll 1968; Murray 1984). The occupation of Czechoslovakia also yielded large stocks of ores and metals that the German Reich was in need of besides gold and foreign currency deposited in the Czech National Bank. Kennedy (1988) is of the belief that Germany, on the eve of World War II, was in such a desperate economic shape as a consequence of its rapid rearmament program that the search for natural resources, precious metals, and foreign currency was the main motivating factor that led to the invasion of Poland following the incorporation of Austria and Czechoslovakia earlier.

In the late twentieth century, the Persian Gulf War also started with Iraq invading Kuwait over oil issues. Though the coalition formed under the auspices of the United States to push Iraq out of Kuwait was rationalized based on stopping aggression and national sovereignty, the protection of access to Middle Eastern oil was most definitely the basis for the formation as the United States, Europe, and Japan were all highly dependent on the oil of the Persian Gulf countries.

The twenty-first century started with a war between the United States and Iraq. Though the main reason given for the invasion of Iraq was over Iraq's presumed possession of weapons of mass destruction (WMD), several commentators such as Harvey (2003), who have tried to analyze *l'histoire événementielle* beyond the surface, have identified the quest to maintain access (and gain control?) to Middle Eastern oil as one of main reasons for the war (see also Greenspan 2007). Beyond the fight for control of Middle Eastern oil, if other commentaries such as the decline of the

United States in the twenty-first century as the sole hegemonic power becomes a reality, then the escalation of conflicts or wars will increase as a consequence of others joisting for hegemonic status while other anti-imperialistic countries, noticing the U.S. decline, attempt to escape the yoke of U.S. imperialism (e.g., see Harvey 2003; Arrighi 2005a, 2005b; Ferguson 2006; Todd 2003).

Such a scenario assumes the analytical perspective of the hegemon at the level of the nation-state, and therefore, if the United States is the hegemon, one can also refer to the American empire as an outcome of the control the United States has over other countries and regions because of its hegemonic status. However, if we shift this to a different level, that of a region, then one can envisage that perhaps what is going on is the decline of one region (the West—European Union and the United States) and the resurgence of another (Asia). Such a view has been suggested by others as well. The work of Andre Gunder Frank (1998), over a decade ago, projected the return of Asia as the next hegemonic region. In a recent work, Niall Ferguson diagnosed the ascent of the East over the course of the last of the twentieth century: "the crisis of the European Empires, the ultimate result of which was the inexorable revival of Asian power and the descent of the West" (2006, lxviii). Despite the fact that Frank and Ferguson come from different ends of the intellectual political spectrum, both concluded with a similar outlook about the recent past and the distant future.

Empire and Decline

With the fall of the former Soviet Union in 1989, the harsh reality emerged on the geopolitical scene that the United States is the only superpower left on the world stage and, hence, the only remaining hegemon in the world system. This development, coupled with the initiatives undertaken by the Bush administration after September 11, 2001, in the area of foreign policy and the subsequent invasion of Iraq, has led to the view by some that the United States is exercising its superpower-hegemonic status or trying to prevent the decline of its hegemonic status. Consequently, the debate has focused on the dynamics of an American empire (Johnson 2000; Harvey 2003; Ali 2003; Ferguson 2004; Todd 2003; Arrighi 2005a, 2005b; Mann 2003; Hardt and Negri 2000).

At the analytical level, according to Hardt and Negri (2000) a new regime determining the global dynamics has emerged. To Hardt and Negri (2000) with the advent of globalization, imperialism (as territoriality) has ended, and the word *empire* should be treated as a concept that defines a web or network operating in a decentralized mode. It is a new regime whose object is the rule of social life in its entirety. Others such as Harvey (2003), Ali (2003), Mann (2003), and Todd (2003), however, based their analyses on the old adage of *empire*, meaning an imperial power, such as America, with its imperialistic hegemonic determination of other nation-states over geopolitical space. Regardless of these different interpretations of empire, the common denominator remains that of power, whether centralized or decentralized, and the ability or inability to exercise it to meet certain economic and geopolitical objectives. It is the exercise of it whether to maintain or prevent the loss of hegemonic status that generates conflicts and resistance. If we follow Hardt and Negri's interpretation of empire as a new regime, social order is derived from several authorities and institutions through flexible principles of regulation whereby centralized and decentralized forms of political rule exist. There is the development of flexible boundaries with mobile labor force characterized by market networks and circulation embedded in hybrid identities and multiple citizenships.

Hybridizing the theoretical conceptions of Hardt and Negri (2000) with those of Harvey (2003) and others, the changing political, economic, and social landscapes of the late twentieth century and those of the twenty-first do resemble the conflicting dynamics of empire whether we agree with the idea of U.S. hegemonic decline (or the decline of the West à la Ferguson 2006) or not. The inability of the United States primarily to achieve its foreign policy objectives in the twenty-first century is seen by some as evidence of hegemonic decline (see Harvey 2003; Arrighi 2005a, 2005b; Todd 2003). Others such as Ferguson (2004) see it as need on the part of the United States to assume its imperial power for self-interest and altruism so that global order can be maintained. What is clear is that regardless of which position you agree with, resistance from the Other (or multitudes) has been and will be the reaction to any corelike initiatives. It has always occurred in history and why should it not be otherwise? Hence, we will witness the continued conflicts, especially during a period of ecological stress, climate changes, and socioeconomic and political turbulences.

If one shifts from nation-state as hegemon to that of region as being hegemonic, conflicts will (have) result(ed) when one region starts to lose its enjoyed hegemonic status. Whether this will increasingly be the case for the rest of the twenty-first century is a big question. If the shift in hegemonic region entails also a shift involving regions that are distinguished by different cultural "socioscapes," then perhaps conflicts might result—that is, if one assumes that differences in culture world-views can generate conflicting situations. Huntington's (1996) and Barber's (1995) arguments that cultural clashes or conflicts defined by traditional cultural practices versus modernistic-globalist orientations have occurred and will likely, in Huntington's case, determine wars and conflicts in the future, thus remaking world order. Ferguson's (2006) argument is that the decline of European empires that started during the beginning of the twentieth century and the revival of Asian societies is the arc that we should follow in understanding changing global circumstances and conflicts. Listen to what he has stated:

> In that sense it seems justifiable to interpret the twentieth century not as the triumph but as the descent of the West, with the Second World War as the decisive turning point. It was a descent, in the sense that the West could never again wield the power it had enjoyed in 1900. (2006, lxix)

If hegemonic decline is the order of the day whether it is single power or that of the West, invariably comparisons need to be made between the current decline and that of Rome 1,700 years ago during the Dark Age of Antiquity. Niall Ferguson's (2004) *Colossus* in a brief manner alluded to some of the characteristics of Rome's Empire that are similar to that of the United States that he assumed the latter to possess. Early beginnings of the United States from thirteen colonies to what it is now covering North America is somewhat similar to Rome's rise starting from an urbanized enclave and, over time through imperial conquest, achieving control of a large land mass. Though in the case of the United States, as Ferguson (2004) has stated, the addition of territories was mostly through land purchases (Louisiana, Texas, Alaska, New Mexico, Arizona, California, Colorado, Utah, and Nevada) beside conquests over Native American populations. Like Rome, in the history of the United States, there was the utilization of disenfranchised slaves in its economy. Such extreme political and labor relationships are further contrasted by the awarding of citizenship to noncitizens who have served in the U.S. military. In a very similar fashion, Rome granted *civis romanus* to foreigners (barbarians) who served in the legions. This granting of citizenship was a pragmatic solution to the recruitment problems of the legion as a consequence of land-tenure changes and taxation crisis, which resulted in a drop in the level of Roman citizens being able to serve in the legions. Furthermore, besides Ferguson (2004), other commentators (such as Murphy 2007) have also noted the similarity in the manner by which the Roman language and culture was adopted and embraced by elites throughout the Empire and in the various parts of the world during Roman times to what has occurred for the United States (in terms of soft power) in the late twentieth to the twenty-first centuries.

Another parallel feature that has to been noted is the size and role of the military in the political economy of the two empires. As I have indicated in *The Recurring Dark Ages*, the Roman Empire's motor force was its legions that initially conquered territories, and from its emplacements in these territories, civic and manufacturing functions powered the economy of the Empire. Throughout the Empire's lifetime, it was the military charges that made up the lion's share of the overall Roman economy. To fund the expanding military budget, Roman emperors from the late third century onward started to debase the coinage of Rome by diluting its silver content. The other route taken was to increase taxes, which was equally as unpopular. With debasement, inflation followed. All these measures to maintain a heightened military posture to address Roman security threats that were coming from the barbarians and Persia in the long run led to tremendous stress economically, politically, and socially for the Roman Empire.

In the case of the United States, it was the U.S. Army in the early periods that saw to the expansion of the country's territorial boundaries in the West and South from the Native Americans and other powers such as Britain, Spain, France, and Mexico. In the late twentieth and early twenty-first centuries, the U.S. military, like its Roman counterparts, has a predominant share of the U.S. national budget. For year 2007, the defense budget was 17 percent of the overall national budget. Deficit financing has also been the avenue sought by the U.S. government to finance its overall budget expenditures. Unlike Rome, the U.S. government did not raise taxes over different administrations from President Ronald Reagan onward other than the administration of President George H. W. Bush to finance its budget deficit. Given this condition, in fact, the only feature that the United States possesses that surpasses other nation-states is its military might and hence its military expenditures, for its share of global gross national product (GNP) has dropped since the World War II, even though it still has the largest share of global GNP (Reus-Smit 2004).

There are other parallels causing upheavals that have not been examined that might not necessarily be just distinguishing elements defining the Roman and U.S. empires, but rather are characteristics depicting social, economic, and political trends during Dark Ages. Migration is one such characteristic that occurred during these periods of stress that was a factor in the final demise of the Roman system.

Incursions and Migrations

"Barbarian" migrations or incursions from the hinterland to core centers over world history for the last five thousand years have been identified (see Chernykh 1992; Toynbee 1962; Teggart 1969). Whether these "eruptions" according to Toynbee (1962) have implications for sedentary societies (i.e., core centers) is relevant to our present discussion. Do these migrations and incursions occur during periods of social, political, and economic stresses that would be pertinent to our understanding of the frequency of upheavals? Notwithstanding the difficulty in gathering the data and classifying the types of migrations or incursions, Modelski and Thompson (1999) have periodized the frequency of incursions of core-hinterland dynamics along a periodization scale they developed to distinguish periods of growth of the world system (concentration) with periods of dispersal of the world system (deconcentration). Preliminary indications suggest that during periods of core concentration there were considerably fewer hinterland incursions than during periods of deconcentration. Modelski and Thompson (1999) confirmed that the period between A.D. 400 and A.D. 700 (identified by Chernykh 1992) was a time of hinterland attacks on the centers, though their periodization is much longer. It stretched from 100 B.C. to A.D. 930. Period-wise, this time sequencing falls during the Dark Age of Antiquity and also the decline of the Western Roman Empire.[7] What also needs to be emphasized is the part played by these hinterland incursions in the process of the breakdown of system configurations during the periods in which they occurred (Modelski and Thompson 1999).

What precipitated these movements of peoples? If one follows Toynbee (1962), the push factor for the mostly nomadic tribes across Eurasia was the change in steppe climate. Greater aridity increased the conflicts among nomads for access to water and pasturage. Others, such as Modelski and Thompson (1999), tend to discount single-factor climatic theories. Clearly, there has not been an agreement on what factors caused the push during these periods of the Dark Ages. Those offering climate change as the push factor have been classified as climate determinists (e.g., Huntington 1924), and intellectual support is most often given to those that offer explanations based on socioeconomic and political grounds. In *The Recurring Dark Ages*, historical evidence was provided to underline the fact that during these periods of distress, changes in climate and ecology punctuated these Dark Age phases, and the Dark Age of Antiquity was no exception.

The Gothic Migrations

The Dark Age of Antiquity was one phase that was characterized by large-scale migrations and incursions that led in the end to having an impact on the decline of the Western Roman Empire. It was a period filled with conflicts, struggles, and compromises, as well as climate changes. Climate shifts occurred in Europe and the Mediterranean as early as the first millennium A.D. Relative dryness predominated until A.D. 400, followed with increasing drought conditions from A.D. 500 (Briffa 1999; Chew 2007; Randsborg 1991; Allen et al. 1996; Lamb 1981). For Africa, relative dryness started much earlier from about 800 B.C. and lasted until A.D. 400, with regular rainfall occurring during the fourth and fifth centuries.

Drought conditions returned from A.D. 600 onward and stayed until A.D. 1250 (Veschuren 2004; Chew 2007). In Asia, China experienced abnormal drought conditions between A.D. 400 and 700. The warm period started from A.D. 1 to 240, followed by a cool period from A.D. 240 to 600/900 (Bao et al. 2002; Ge et al. 2003). Tan and Liu (2003) and Brown (2001), however, reported a warm period for China from A.D. 600 to 800.

The system-wide scale of climate changes occurring over long duration generates stresses on physical and ecological landscapes, including human communities in their reproduction of socioeconomic life. Human groups that are nomadic in nature and others that are focused more on agricultural pursuits either in crop cultivation or animal husbandry are more likely to be impacted by changes in the physical and ecological landscapes. Especially for the nomads, their natural environments dictate their socioeconomic activity. Usually domiciling in grassland plains devoid of fertile soil and sufficient rainfall, their pursuits are overwhelmingly focused on herding a range of animals. During the Age of Antiquity, most often such groups with these socioeconomic characteristics and lifestyles living outside the Roman Empire were labeled as "barbarians" by the Romans for they were deemed to be less socially, economically, and politically developed in terms of organizational complexity. Cullen (2007) has suggested that the word *barbarian* connotes incomprehensibility to the Romans. According to him (2007, 133) the word is "generally said to be an onomatopoeic term that originated with the Greeks, to whose ears the speech of outlanders sounded like a meaningless 'bar-bar-bar.'" Furthermore, for the rational Roman, these tribal groups were labeled "barbarians" because they let their bodies rule their minds instead of the other way around, which is supposed to depict the human—that is, the Roman (Heather 2006). In the postmodern sense, or our contemporary period, they are the depicted as the "Other."

In terms of numerical totals and diversity and in linguistic competence, the Others, when we examine the history of this period, are quite diverse and numerous in tribal groups. For example, during the first century A.D., beyond the Roman frontier in western and eastern Europe, Germanic linguistic groups, of which there were many, peopled most of central and northern Europe. Celtic tribes resided in eastern lands beyond the River Vistula. The Sarmatian nomads roamed the Great Hungarian Plain, and Dacian communities settled around the Carpathians. As indicated in *The Recurring Dark Ages* and in the previous pages, the relationships between Rome and these barbarian communities were of various orders: some were treated like client states while others were subjugated under Roman imperial rule where at times spurts of skirmishes between these barbarians and Rome occurred (Wolfram 1988). Those which had client state status enjoyed unequal relations as well (see Heather 1991). They were subjected to the provision of recruits for Rome's armies and had their political affairs interfered with also.[8] In turn, the Romans would provide annual payments and land grants according to the treaties that were signed.

During the early Roman period, it was a fragmented world with over fifty sociopolitical units surrounding and facing the Roman frontier (Heather 1996). By mid-third century A.D. the Franks, the Alamanni, the Saxons, and the Burgundians replaced the earlier tribal groups such as the Cherusci, the Chatti, the Bructeri, and the Ampsivarii on the Rhine frontier. Further south, the Sarmatians continued their presence.

To understand border incursions and migrations they need to be placed within the structural changes that occurred from the first to the third centuries A.D. During this period, Germania underwent massive population increases, economic expansion, and social and political differentiation. The expansive growth trajectory is exhibited by the grave goods excavated, and the various literary accounts of the period (Heather and Matthews 1991; Heather 1991, 1996; Wolfram 1988). Because of such transformations, any system stress of ecological, climatological, socioeconomic, and political nature would greatly impact these Germanic tribal groups. The sequence of events would be as follows: those Germanic communities that were Roman client states, besides having to live under Roman imperial rules, often also faced invasive threats from peripheral tribal groups who were located further away from the Roman frontier. Pressure on the peripheral communities residing in the hinterland areas away from the Roman frontier would come from changing climatological conditions, devastated landscapes, and socioeconomic downturns. Their natural reactions to these conditions would be to migrate or invade the lands of the Germanic client states that bordered the Roman Empire's frontier. These invasive movements would thus push the leadership and population of the Germanic client states that bordered the Roman frontier to seek safety within the borders of the Roman Empire.

The specific causes (climatological change or other socioeconomic or political reasons) that conduced these peripheral groups to encroach on the lands of the Germanic client states for each specific migration are hard to tease out. But we do know that these occurrences were quite frequent from the second century A.D. onward when the environment had periods of ecological devastation, climatological changes, and socioeconomic and political unrests as outlined in *The Recurring Dark Ages*. Literary accounts have also identified several such occasions from A.D. 376 to 450 when these migrations or incursions took place.

One of the most notable occasions in the history of the Roman Empire was the initial migration of two tribal Gothic communities (the Tervingi and the Greuthungi) that started around A.D. 376 (Pirenne 1992). The event goes as follows: Around A.D. 376, approximately two hundred thousand

Goths from the Tervingi and the Greuthungi tribes appeared on the north bank of the River Danube seeking to cross over to the territorial boundaries of the Roman Empire that bordered the south bank. This massive migration was instigated by the marauding Huns from central Eurasia encroaching and raiding Gothic lands north of the River Danube (Pirenne 1992). Envoys from the Goths were sent to negotiate with the Roman emperor Valens for permission to resettle within the Roman Empire. However, permission was only provided by the emperor for the Tervingi to cross over. The Greuthungi tribe accompanying the Tervingi was denied entry at first. The denial to the Greuthungi according to literary sources was based on the fact that the Tervingi had closer relations with Rome and that they were a "better known quantity" according to Heather (1991, 131). However, with the limited number of Roman troops available to guard the Tervingi for their move to Marcianople following their entry to the Roman Empire, the Greuthungi slipped over when the Roman legion guarding the frontier was reassigned to watch over the Tervingi relocation and Roman river patrols were curtailed (Wolfram 1988).

Using classic migration push-pull factors to explain the movement of the Gothic tribes, one can surmise that for these tribes, the pull factor for migrating into the Roman Empire was to transport to an area of safety from the Huns and to enjoy the economic benefits of the Roman Empire. Of course, the push factor would be to escape from the invading Huns. Having accounted for the movements of the Goths, what about the Huns in terms of the conditions that prompted them to move out of their homeland? The reasons for such a Hunnic migration are not very clear. A number of reasons or causes have been put forth. Population pressure has been raised as a factor. The changing climatological condition leading to drought conditions on the steppes has also been brought up as a possibility. Of course, the usual standard socioeconomic attraction of richer grazing lands to the western end of the Eurasian steppes cannot also be discounted. Whatever the reasons, the environment seems to be one of the major factors that underlie the Hunnic migration/incursion in the fourth century.

The above incidence of A.D. 376 reveals to us the migratory tendencies of tribal groups during the Dark Age of Antiquity. By no means were these movements the only occasions whereby we witness these migratory flows. Throughout the Dark Age of Antiquity there were numerous occasions whereby such movements took place. For example, later in the period around A.D. 405–406, the Gothic king Raidagaisus led a multitribal force and invaded Italy. He was defeated but the event shows the constant incursion of the migratory barbarian tribal groups. Four months later, following the death of Radagaisus, an army of Vandals, Alans, and Suevii crossed the Rhine and marched into Roman Gaul. In this case also, one of the tribal groups, the Alans, was trying to escape from the marauding Huns.

Besides the Goths, the Huns also invaded the Roman Empire. Two years later, around A.D. 408, the Hunnic leader Uldin led an army and seized Castra Martis. This was not the first occasion of Uldin leading a Hunnic army into the Roman Empire. He had done so on numerous occasions. To avoid the repetition of listing the various incursions into the Roman Empire, let us focus instead on the effect these migratory flows have on system stability.

The Gothic migration-incursion that started in A.D. 376 is a good exemplar of the destabilization migratory flows have on system stability and reproduction. Noting that perhaps environmental and climatological changes forced the Huns to encroach on Gothic lands north of the Danube and, in turn, forced the Goths to flee to the Roman Empire, Gothic resettlement on Roman lands had consequences for the stability of the Roman Empire.

Gothic resettlement within the Roman Empire was by no means uneventful. It was fraught with conflicts. Right from the start, both sides—the Romans and the Goths—were in constant tension. Social and political dynamics within both camps generated strains on the resettlement agreement. From the beginning, following the entry of both Gothic tribes, tensions were already generated by the Romans through various attempts to subjugate the Goths. On the part of the Goths, they were also trying to maximize their own advantage. Thus, a series of armed conflicts resulted with severe losses on the part of the Romans. One such loss occurred when the Roman legions under the Roman emperor Valens were decisively defeated in a battle at Thrace in A.D. 378. This loss is significant as it was the first time a barbarian Gothic army had defeated the Roman military. The victory caused tremendous destabilization of the Eastern Roman Empire as it gave the Goths control of the Balkans. With this, the path to Constantinople lay open. Unfortunately, the Goths never made it to Constantinople as the Roman legions of the Western Roman Empire arrived, and over several campaigns they drove the Goths back to Thrace. A peace treaty was signed in A.D. 382.

This treaty was in force until the death of Emperor Theodosius in A.D. 395. The resulting power vacuum prompted the Goths under Alaric to rebel as they sought further accommodation from the Romans. Even with the new arrangements and following the death of the Roman general Stilicho, Alaric moved on Rome in early A.D. 409 with Goth deserters from the Roman legions. Following failed negotiations with the western Roman emperor Honorius, Alaric and his Gothic army entered and sacked Rome. The Gothic entry and ransacking of Rome shook the Roman world to its foundations (Heather 2006). From the Roman point of view, the Other has succeeded in attacking its homeland. With Alaric's untimely death, social and military upheavals continued with the Gothic army moving up and down the Italian boot and finally making it into Gaul. By A.D. 417, a new political unit, the Visigoths, was formed in the west comprising of the Tervingi and the Greuthungi tribes along with followers of the Gothic king Radagaisus. After 418, the Roman Empire was forced to recognize the leadership of the Gothic king in Gaul. With this, the acceptance of an autonomous Gothic settlement within the territorial boundaries of the Roman Empire was sealed. Similarly in the east, two Gothic groups, Amal-led Goths and Goths of Theoderic Strabo, formed the Ostrogoths circa A.D. 450 that later challenged the Eastern Roman Empire between A.D. 474 and 489.

Gothic migratory incursion by no means was the only destabilizer of the Roman Empire. The Huns' role in A.D. 405–408 should also be seen as the other destabilizing force on the Roman Empire. Initially, the Hunnic advance in the late fourth century caused a crisis in Germania. It also generated alarm in the Eastern Roman Empire. Throughout the early parts of the fifth century, the Huns were a major threat to the stability of the Roman Empire whereby its move across Europe would displace populations and also the authority of Roman rule in the conquered territories. By the mid-fifth century, the threat reached its apogee with the arrival of Attila the Hun. Attila's empire was not centered in the lower Danube region like Uldin, but was located in the Greater Hungarian Plain west of the Carpathians (Heather 1996, 2006; Thompson 1996). In two successive campaigns, one in A.D. 443 and the other in A.D. 447, Attila won major victories over the Roman military and threatened the capital of the Eastern Roman Empire, Constantinople.

What were the causes for such invasive practices? Besides population pressure and changing climatological conditions in the steppes, most historians have suggested that it was because the Romans abrogated earlier treaties requiring annual payments of gold to the Huns. The arrears amounted to 1,400–2,100 pounds of gold per annum from A.D. 442 onward (Bell-Fialkoff 2000, 224). Such payments to be made annually was a drain on Roman resources, especially at a time when the gold mines of its empire, which had been exploited for quite some time, were running dry.

Having scored a victory in the Eastern Roman Empire, Attila turned his Hunnic army on the western portion of the Roman Empire: Gaul (Thompson 1996; Heather 1996). Orleans was its first stop where the Hunnic army laid siege to the city peopled by Alans in the service of the Roman legion. The conflict stalemate that followed directed Attila to switch his attention to Italy where he marched his army across the Alpine passes. Aquileia was taken first and west across the River Po's plains the Hunnic army marched. In its path, the cities of Padua, Mantua, Vicentia, Verona, Brescia, and Bergamo were captured and looted. Following entreaties from the Pope not to invade Rome, and experiencing food shortages and diseases along with pressure from the legions of the Eastern Roman Empire on the Hunnic Empire in the East, Attila's army turned back to the Hungarian plain and central Europe. Attila's intention of further attacks on the Roman Empire did not stop with this retreat. However, on the eve of starting a new campaign in A.D. 453, he passed away quite suddenly following a wedding feast after taking a new wife. If one traces this initial impetus of Hunnic migration as one of changing climate and ecological conditions in the steppes of Central Eurasia, a sequence of events and history resulted indirectly affecting the reproduction of the Roman Empire.

The Hunnic migrations that started from A.D. 376 onward in several phases (A.D. 376–380, A.D. 405–408, A.D. 440–450s) had tremendous consequences for the stability of the Roman Empire (Thompson 1996; Heather 1996). Though the Huns did not invade the Roman Empire in A.D.376, their incursions into Gothic homelands forced the Goths to seek sanctuary within the confines of the Roman Empire. This resettlement of the Goths as discussed had tremendous repercussions financially and politically for the Roman Empire. As we have seen, in the end the Goths, through several social self-formations were no longer permanent social-political enclaves in southwestern Gaul and in North Africa of the Roman Empire. The mid-fifth century incursions (A.D. 440s onward) of the Roman Empire by Attila generated other stresses on the reproduction of the Roman Empire. In the east, the Roman Empire suffered a tremendous level of damages in its territories, especially in the Balkans following Attila's raids. Population displacements and social disruptions of the elite structure placed further tension on the social fabric of life in the Eastern Roman Empire. The displacement caused pressure on tax collections as the population base dwindled and/or shifted. In turn, there was tremendous pressure to meet tax obligations. From A.D. 439 to 441, the tax revenues from this region turned to a trickle. This had tremendous consequences for the Roman treasury, which already had to provide annual gold payments to the Huns.

Besides these direct consequences of the various phases of the Hunnic incursions, other indirect effects were also felt. In the other parts of the Roman Empire, because of Attila's threat to the armies of the Eastern Roman Empire, there was little defense protection afforded to Spain and North Africa as the Roman legions were redeployed to the Balkans and the east to meet the threat of Attila's army. With this military security vacuum, North Africa was eventually conquered by the Vandals. In Spain, most of the territory was abandoned and left to be shared by the Vandals, Alans, and Suevi. Roman Britain also followed the same fate. Incursions by Saxons from northern Gaul facilitated the Roman departure from this remote part of the Roman Empire. By A.D. 452, the indirect effects of the Hunnic invasion of the Roman Empire had resulted in the loss of Britain, Spain, parts of North Africa, and southwestern and southeastern Gaul (ceded to various Goth tribal groups such as the Visigoths, etc.).

Over the long term, environmental conditions and weather patterns affected the reproductive lifestyles of tribal groups. Though not clearly and easily distinguishable because of the common propensity of historians, archaeologists, and other social scientists to gaze toward socioeconomic (access to land and employment) and political factors as the instigators of migrations and incursions, environmental factors in view of our discussion should not be dismissed so quickly. If "history is the teacher of life" (Schäfer 2007), can we then consider the pattern of migrations and incursions that occurred in the Dark Ages (Bronze and Antiquity Dark Ages) as indicators of system tendencies? If so, then we can assess whether the large-scale migratory patterns that have and are occurring in the late twentieth to the twenty-first centuries are similar system tendencies reflecting the possibility of Dark Age conditions or are the threat of an impending one.

Social Worlds in Motion: The Nineteenth and Twentieth Centuries

Movements of people have continued throughout the nineteenth and twentieth centuries. Over this time horizon, the large-scale movements have been within kingdoms, principalities, and empires; immigration to lands that were colonized or conquered that later became migrant states; or migration to fully developed states that have historical identities. For the migrations of the late twentieth and early twenty-first centuries, the large-scale movements have mostly been ones that occurred between the reconfigured political frameworks of nation-state. For the latter period (i.e., the late twentieth century onward), the pattern of migration has been from the peripheral areas to the core zone of the world system and within peripheral zones, though the former surpasses the latter. The travel routings of these movements have been circuitous occurring within regions and between regions of the world (e.g., see Castles and Miller 2003; International Organization for Migration 2003).

Regardless of migration over the expanse of geographic space, some of the push factors for the movements during the nineteenth and twentieth centuries bear certain similarities to the fourth and fifth centuries' patterns in Roman Europe. Notwithstanding the commonly identified push factors that are socioeconomic and political in origin that cause migrations, climate change and landscape degradation have also been identified as causes for large-scale movement of peoples during the nineteenth and twentieth centuries (see Brown 2006; Castles and Miller 2003; Wang

1997; Zolberg and Benda 2001; United Nations World Economic and Social Survey Part 2 2004). One such movement of people, the migration of the Irish in the mid-nineteenth century (1845–1849) to North America as a consequence of crop failure is an excellent historical example in which all factors ranging from socioeconomic and political to climate changes have been put forth to account for its occurrence (Ó Gráda 1999; Kinealy 1997). Numerous studies have been done to account for the socioeconomic and political causes leading to Irish migration (e.g., Chew 1992; Ó Gráda 1999; Kinealy 1997; Bourke 1993). These studies suggest the following as the primary specific socioeconomic factors leading to the crisis and the subsequent disaster:

1. Ireland's colonial relationship with Britain shaping a land tenure system characterized by absentee landlords,
2. farm sizes that were too small for the reproduction of the Irish family, including inheritance rights,
3. crop overspecialization as a consequence of harvest yields,
4. tax obligations of tenant farmers, and
5. increasing prices of food crops due to elevated demand to feed the industrial workers in England, hence resulting in the inflation of land prices thus providing an opportunity for landlords to increase land tenure rental costs.

All these specific contextual conditions structured socioeconomic and political organizations and thus increased the vulnerability of the Irish economy and social milieu to famine conditions.

In the case of the environmental conditions that precipitated crop failure, it has been suggested that the climatological circumstances during the crisis years had a tremendous impact on the harvest yield of the Irish potato crop. Through changing weather patterns conditions were set for fungus spores to attack the potato bulbs, causing blight (Ó Gráda 1999; Kinealy 1997). The explanation is as follows: In 1846, bad weather delayed the spring planting, and with a low rainfall in the early parts of the summer, this delayed the growth of the tubers. Heavy rains arrived in late summer causing the spores of the fungi (*Phytophthora infestans*) to be washed into the soil, resulting in the infestation of the potato bulbs. Cold winter conditions in 1846–1848 made matters even worse. The outcome was a decrease in the harvest yield of the potato per acre in Ireland. Historically, it has been noted that Ireland was affected the most in comparison to other places in Europe such as France and Belgium. Scotland as well suffered a similar drop in potato crop yields. However, this decline in harvest yield did not impact Scotland as greatly because its agricultural economy did not rely as heavily on the potato for each household's daily caloric intake in comparison to the Irish household. Hence, when the potato blight hit Irish agriculture, the outcome was devastating because of Irish dependency on the potato as a basic staple food and the socioeconomic framework in place (see Chew 1992). What resulted were famine conditions.

Furthermore, the cold weather made the Irish peasant household even more vulnerable to diseases. The lowered temperatures made working outdoors in the damp, cold conditions unhealthy. The weather changes made famine conditions even worse for the tenant farmers and their families.

The outcome was migration, first to England. Caught with such a massive influx of people, English political authorities passed a series of measures to address the crisis. The final resolution to the crisis was government-subsidized passages to North America. More than 1 million Irish migrated to England and North America over the course of the famine. Over the long term, from the start of the famine to the end of the nineteenth century, approximately 2.811 million Irish left northern and southern Ireland (Chew 1992, 49).

Such a massive movement of people of course had certain consequences for the overall political economy of England and North America. The migrants provided the necessary labor force for various activities such as farming, lumbering, mining, and other industrial activities of North America. Their activities not only facilitated the accumulation of capital on both sides of the Atlantic, they also intensified the Nature and Culture relations through the transformation of the landscape via their support of economic activities such as lumbering, mining, and agriculture. The presence of the Irish also engendered cultural discrimination, and Irish ghettos appeared in the urbanized areas of the northeastern seaboard of America.

By no means were such movements of people conduced by climate and environmental conditions over by the end of the nineteenth century. There were numerous occasions when the "environmental refugee" appeared as we progressed through the twentieth century (El-Hinnawi 1985; Hong et al. 2004; Brown 2006). It has been estimated that there were at least 25 million environmental refugees in the world by 1995 (International Organization for Migration 2000; Myers and Kent 1995; Lohrmann 2000). Myers (1997) has even suggested that more than 200 million people could be affected by the changing environmental conditions. This does not mean that all those that are impacted do move because if the estimate of 25 million environmental refugees is on the mark, then the statistics published by the United Nations Refugee Agency (UNHCR) need to be revised as the total number of refugees assisted by the UNHCR was

26,103,200 (UNHCR 1995). This discrepancy can be due to a number of reasons, such as the definition of an environmental refugee, undercounting, or the difficulty in pinpointing the various causal factors (such as conflicts that result over natural resources) that lead to migration and/or refugee status. The global count for total number of refugees for 2006 that claimed this status and were assisted by UNHCR is 32,861,500 (UNHCR 2007). The proportion of this pool of persons impacted by changing environmental conditions is not available. However, the number of persons estimated to be impacted by natural disasters has been provided by the International Federation of Red Cross and Red Crescent. According to this agency and UNHCR (2006), by 2006 the total number of persons impacted by natural disasters had tripled over the last decade to about 2 billion persons with an accumulated impact of an average of 211 million persons affected directly each year.

To be more specific, these environmental refugees can be produced by varying conditions such as land degradation (through deforestation and intensive farming), leading to the relocation of the farming and rural population to urbanized areas; climate changes, such as droughts or excessive rainfall, leading to crop failures; the process of desertification; water scarcity; or the sudden occurrences of natural disasters forcing evacuations (e.g., see McLeman and Smit 2003; Brown 2006). There have been studies drawing this link but mostly at the micro and regional levels (see Myers and Kent 1995; Brown 2006; Black 1998). Changing environmental conditions such as drought and land deterioration (deforestation, soil erosion, urbanization, etc.) especially in Africa, Central America, South Asia, and Southeast Asia (the Philippines and Indonesia) contributed to the availability of migrants willing to move (Black 1998; Brown 2006; Homer-Dixon and Blitt 1998; Myers and Kent 1995). Degraded and polluted landscapes in Eastern and Central Europe also contributed to the outflow of Eastern and Central Europeans following the collapse of the former Soviet Union in 1989.

Specific country studies on Haiti, Mexico, and the Philippines have shown the linkages between environmental degradation (deforestation, soil erosion, water scarcity) leading to out migration from rural areas (Myers and Kent 1995). In the case of Mexico, which is a major source of legal and illegal migrants to the United States, the link between environmental rundown conditions and migratory movement has been established (Myers and Kent 1995). Mexico's developmental strategies have been rapid industrialization, export agricultural crop production, rapid urbanization, and shifts in the labor-force formation. All these strategic processes have ignored the long-term viability of the environment. With rainfall being sparse in some regions and the need to import food as a consequence of population increases, such stresses have added to increases in poverty levels and reduced the resiliency of the social fabric to crises conditions. Complicating this situation is that four-fifths of the surface water is located in the coastal regions of Mexico whereas three-quarters of the Mexican population and its agricultural production lands are in the semi-arid central highlands (Myers and Kent 1995; Liverman 1992). Given these conditions, 3,250 km^2 of cropland and potentially productive lands are lost to desertification and water shortages each year. Salinization of the crop lands is also a major catastrophe. This has affected 52,000 km^2 and another 10,000 km^2 are threatened, leading to a loss in harvest yield toward the end of the twentieth century. Adding to this environmental deterioration is the continued deforestation that has also its consequences. All these conditions add to the poverty levels of the rural peoples who are already at a disadvantage in view of the low levels of land ownership. The net result has been an immigration to the cities, which has not resolved the problem. Such inflows further add to the stress that urban areas have on natural resources, including water supply, which already are at their limits. These stressful environmental conditions leading to migratory movement were recognized by the government of Mexico whereby it stated to the 1994 International Conference on Population and Development that "the exhaustion of natural resources and the destruction of ecosystems are the cause of unwanted population displacements" (Myers and Kent 1995, 90).

By no means is the situation occurring in Mexico unique. It can be found in almost every part of the world albeit with some specific environmental differences in certain cases. On the whole though, the general contours are similar and the linkage is there (Brown 2006; Myers and Kent 1995). To this extent, we find Hammer (2004, 232) reporting of a substantial number of environmental refugees affected by environmental disasters for West African countries over the course of a quarter of the last century (see table 3.1).

If we extend changing environmental conditions beyond climate changes and natural disasters, other environmental circumstances can also give rise to the movement of population. The impact of development projects leading to a changed habitat, such as dam building for example, can lead to dislocation of people from their homes. The most prominent examples are the building of the Sardar Sarovar Dam in India and the Three Gorges Dam in China (Stein 1998; Xu et al. 1996; Hong et al. 2004). In terms of number of people affected by such development-style projects, Shah (1995, 574) has estimated that for India a total of 20 million persons have been displaced by development projects centered on dam building and irrigation. Industrial accidents such as the Bhopal disaster or the Chernobyl debacle can also provide potential environmental refugees. Wars and armed conflicts leading to environmental degradation can also lead to dislocation (Afolayan 2001).

Table 3.1. Environmental Events in West Africa With More Than 100,000 Refugees, 1973–1999

Year	*State*	*Number of Environmental Refugees*	*Type of Disaster*
1973	Burkina Faso	300,000	Drought
1973	Mali	300,000	Drought
1973	Mauritania	100,000	Drought
1973	Niger	110,000	Drought
1982	Benin	500,000	Flood
1985	Burkina Faso	222,000	Drought
1985	Mali	200,000	Drought
1985	Niger	1,000,000	Drought
1985	Benin	250,000	Flood
1986	Mauritania	220,000	Drought
1988	Nigeria	200,000	Flood
1988	Benin	100,000	Flood
1994	Nigeria	400,000	Flood
1995	Ghana	200,000	Flood
1996	Sierra Leone	200,000	Flood
1998	Nigeria	100,000	Flood
1999	Ghana	144,000	Flood

Source: Hammer 2004, 232.

Notwithstanding the above, anthropogenic causes can also engender changed environmental conditions, leading to migration. In other words, socioeconomic and political situations on certain occasions can lead to environmental-induced displacements. For example, the continued increase in rural populations in the periphery of the world system most often leads to situations of environmental degradation, depletion of natural resources such as deforestation, or impacts on watersheds through extensive irrigation, thus transforming the overall natural environment. Similar situations can also result because of the increase in urbanization, which has the consequence of drawing on the resources of rural areas and often encroach spatially on them. Sub-Saharan Africa exemplifies such circumstances that have led to such migratory patterns. It has been estimated that at least 50 percent of the poor live in such environmentally distressed conditions leading to migrations initially to the urbanized areas and, at times, onward to the core zone of the world system. In addition to these factors, drought and desertification have also been factors in the displacement of the population in Sub-Saharan Africa especially in Rwanda, South Africa, and the Sahel (Adepojo 1995; Hammer 2004). Besides Sub-Saharan Africa, other areas that have experienced these types of environmentally distressed conditions are Central America (El Salvador, Haiti, Mexico), and Asia (the Philippines, India, Bangladesh, Pakistan, China)

(Brown 2006; Homer-Dixon and Blitt 1998; El-Hinnawi 1985; Amacher 1998; Hong et al. 2004).

Given the above, there has been a high volume of migration flows in the 1990s and beyond, though not all these migrants are environmental refugees. For example, the number of international migrants for North America has tripled between 1960 and 2000, going from 13 million to 41 million (UN World Economic and Social Survey 2004). The United States leads the North American countries, including Australia and New Zealand, in terms of the number of migrants.

This pattern is also repeated for Europe, excluding the former USSR. Over the same period (1960–2000), the volume of migration rose from 14 million to 33 million. In Europe, the main receiving countries over this period are Germany, France, United Kingdom (UK), and Italy. Besides these influxes, the Gulf Cooperation Council member countries have also been receiving a large number of migrants and guest workers. From a total of 228,000 persons in 1960, the number of migrants has reached a total of 9.63 million persons. The largest concentration and increases have been for Kuwait, Saudi Arabia, and the United Arab Emirates. Together they account for 86 percent of the total by year 2000 (UN World Economic and Social Survey 2004).

Given this history of population flows, several specific migratory movements over the course of the late twentieth century and the new millennium have shaped certain internal dynamics of the receiving countries, in particular, the legal and illegal migration of Mexicans and Central Americans into the United States, the active recruitment of guest workers by European countries (especially by Germany during the 1970s), the migration to Western Europe and the UK of people from its former colonies, and the movement of guest workers from South Asia and Southeast Asia within Asia and to the Gulf Cooperation Council member countries.

Prior to the oil crisis of 1973–1974, the West European economies were experiencing labor shortages and thus were receptive to the import of guest workers to make up the deficit (Castles and Miller 2003). France, Netherlands, and the Federal Republic of Germany approved the formal entry of migrants especially from southern Europe, Turkey, and North Africa. For Britain, the net inflow came from its former colonies such as Ireland, the Caribbean, South Asia (Pakistan and India), and Africa. By 1970, France had over 2 million foreign workers including 690,000 dependents. Germany, with its booming economy prior to 1973–1974, had the most elaborate system of controls. By 1973 its foreign guest-worker population had reached a total of 2.6 million. This program of active recruitment came to a halt with the recession of the 1970s. Migratory movements did not stop however. In the second half of the 1980s, there was a resurgence of migration to Western Europe. By this time, there were also legal and illegal migrants from Africa, Asia, Latin America, and Eastern Europe (OECD 2000). As a result, for France the total amount of foreign residents reached a total of 3.2 million by 1999. This amounted to about 5.6 percent of its total population. Germany had a population of 7.3 million foreign residents, making up 9 percent of its total population. The UK had 2.2 million foreign residents. This came up to about 4 percent of its population. With the fall of the Soviet Union in 1989, there was a further influx of migrants from the former Eastern Bloc countries into the European Union. Approximately 2.4 million migrated between 1990 and 1997 from Central and Eastern Europe.

By no means is the European Union the only center of legal and illegal migration. The Gulf Cooperation Council member countries and the booming newly industrializing countries of Asia (Malaysia, Singapore, Taiwan, Hong Kong, and South Korea) and Japan were also magnet points for legal and illegal migrants. Starting with the end of the Vietnam War, which led to massive outflow to the West and Southeast Asia, the main sending countries to the Middle East and Southeast Asia were Pakistan, India, Sri Lanka, Indonesia, Thailand, and the Philippines. In addition to these areas, Europe, North America, Australia, and New Zealand were also migrant destinations of choice.

By 1985, there were 3.2 million Asian workers in the Gulf countries. This influx increased to almost 1 million workers leaving South Asia for the Gulf by the late 1990s. According to the International Organization for Migration (2000, 109) in 1994, the outflows were as follows to the Gulf: India (405,000), Bangladesh (133,000), Pakistan (114,000), and Sri Lanka (111,000). In addition to nationals of these countries, the Gulf countries also received worker migrants from the Middle East, especially from Egypt, Palestine, Yemen, Lebanon, Sudan, and Jordan.

Within Asia, there was also regional migration. Countries such as Indonesia, Thailand, and the Philippines, with stagnating economies and rising poverty, supplied labor for newly industrializing economies including Japan. Japan, with its booming economy throughout the 1980s–1990s, imported migrant workers. Between 1975 and 2001, the number of legal resident workers increased from 750,000 to 1.8 million (United Nations World Economic and Social Survey 2004). Taiwan also exhibited such worker migrant trajectory. From 98,000 in 1993, the number of worker migrants rose to 316,000 by 2000. Malaysia also has a long history of importing the migrant worker. By 1993, there were 1.2 million migrant workers in Malaysia with this total reaching 1.4 million by 2000. Hong Kong and Singapore also received migrant workers especially in the service sector of their economies. Singapore, for example, in 1993 had over 81,000 foreign domestic servants with the majority coming from the Philippines.

In terms of labor profile, a predominant overrepresentation of the female worker emerged. From 1984 to 1994, over two-thirds of Indonesian migrants were female. This was also the case for worker migrants from the Philippines as the number of female workers rose from 50 percent in 1992 to 61 percent in 1998 (Castles and Miller 2003). On the whole migrant workers, irrespective of gender, are concentrated in the construction, service, and labor-intensive production sectors of the economies of these receiving countries.

By far one of the most extensive and large-scale migrations occurred in North America, and it is still in progress. Total immigration to the United States rose from 4.5 million people in 1971 to 1980, to 7.3 million from 1981 to 1990, and to 9.1 million from 1991 to 2000. In addition to this, the number of undocumented or illegal migrants was about 9 million in 2000, and the current estimate is about 12 million. Countries sending migrants to the United States were Central America and Mexico, the Caribbean, China, India, and Europe. The migration of Mexicans into the United States started as early as the 1940s with the Bracero Program. The flow since then has not stopped, and throughout the 1980s to the new millennium the rate has increased with illegal migrants crossing the U.S.-Mexican border each year. Central American migration started to gain momentum during the unrest in Central America in the 1980s. Besides these countries, migration from Asia, especially from China and India, along with those countries from the Caribbean have also been main sources of legal and illegal migrants. For year 2000, according to the International Organization for Migration (2003), the main regions and countries in terms of legal migrants to the United States were Asia and the Pacific (265,400), Mexico (173, 909), Central America and the Caribbean (154,686), Europe (132,480), South America (56,074), China (45,652), Africa and the Middle East (44, 731), Canada (16,210), and Other (1,181).

Canada also has an active immigration program. Though smaller in volume, the sending regions and countries are as follows: Asia and the Pacific (120,491), Europe (42,875), Africa and the Middle East (40,779), China (36,718), South and Central America (16,939), the United States (5,809), and the United Kingdom (4,648).

With the above volume of migratory flows, what can we expect in the future in terms of the total number of possible of global environmental migrants with the expected global warming and deteriorating environmental conditions? For year 2025, Myers and Kent (1995) have projected that at least 225 million persons will be affected by increasing desertification. Forecasting an expected number of impacted individuals is difficult. A forecasted projection of 150 million environmental refugees has been provided by Myers (1993) by the year 2050. Given such a potential level of environmental migrants what is clear is that global social instability will increase.

Migration, Cultural Upheavals, and Globalization

During Roman times, conquered persons and others who moved to live within the Roman Empire were awarded *civis Romanus*. As citizens of Rome they enjoyed the same benefits and rights as those living within the Roman Empire. Adoption of Roman lifestyles and religion was the standard practice of the day as these barbarians found the Roman way of life and civilization practices "progressive" and enhancing. Despite such assimilation tendencies the disruptive pressure on the Roman Empire from these integrated barbarians did not subside as the Western Roman Empire was under constant attack within and without toward its end. If history continues to be our teacher and knowing the historical challenges and social ruptures that occurred during the decline of the Western Roman Empire, what can we anticipate in terms of social and cultural stability for this millennium as a result of the inflow of migrants from the south to the north that has gained momentum during the late twentieth century?

For the late twentieth century for some receiving countries there were already signs of cultural disruptions developing primarily as a reaction to the inflow of migrants or guest workers whose ethnicity or historical cultural past was not similar to the historical identity of the receiving country. In this regard, France, Germany, and the UK have had their incidences of cultural conflicts (see Castles and Miller 2003; Castles 2000). Racist acts of violence have been perpetrated on mostly visible legal and illegal migrants and guest workers. Such a volatile environment has led to calls for ending immigration. As a result, political mobilization to end immigration has occurred through campaigns of political parties such as the Front National in France (led by Le Pen), the National Front, and the British Movement in Britain. On the whole to date, such calls have come from right wing parties and neo-Nazi groups. The latter have also appeared in these countries challenging the rights of visible minority migrants and guest workers. In Germany, however, the violence has not only been leveled at visible minorities, such as the Turkish population, but it also has been transferred to former residents of East Germany following the fall of the Berlin Wall and German reunification in 1990. With its large influx of illegal migrants from Mexico, the United States is also showing signs of racist discriminatory acts and state policies against them. The founding of various vigilante groups to assist in the patrolling of the U.S. border with Mexico is a clear sign of the increasing discriminatory tendencies against visible minority illegal migrants. Furthermore, passage of various laws at state and federal levels denying social and health programs to illegal immigrants further adds to this discriminatory tendency.

Whereas the actions identified above are directed by the natives and to certain extent local state legislatures on the illegal migrants and guest workers, the emergence of fundamentalist Islamic groups has added a different volatility to the equation. In this case, the growing sense of cultural-religious identity has led some visible minorities, whether they are native born or legal or illegal migrants, to voice their social and political rights. Numerically their mobilization can gain traction in the future. As of 2002, there were about 15 million Muslims in Western Europe, of which over 4 million were in France. In France, the challenge to the indigenous French social life has been raised on certain occasions, from the wearing of headscarves by Muslim students to the call for the establishment of religious schools. These demands for religious fundamental rights have made the sociopolitical atmosphere quite charged. With the attacks in New York on September 11, 2001, and the worry of terrorism's spread, these fundamentalist Islamic movements and ideas have added further anxiety to the cultural-security equation within the United States. Whether this is true or not does not really matter. It is the associative symbolic representations along with the rhetoric that have made the situation more charged.

How does one make sense of these tendencies in terms of social stability of the system and what role does cultural security play in adding pressure to the transformation of the system? As we have seen in the case of Rome, assimilation practices did not further the social and political stability of the empire. At least as far as we can see, it did not diffuse the social and political tensions existing then, as the various tribal groups continued to pressure the western part of the Roman Empire (Heather 2006). In the current context whereby the situations are more complex and furthermore quite different in each country, the calculus of cultural reactions and security is even more complicated to project. Explanations that have been based on either economic or social grounds for the turmoil we are witnessing are quite limited in their explanatory power. Rather, current social and political conflicts have to be understood within a set of structural tendencies and, as well, a consequence of cultural politics and dynamics.

The structural component I am referring to covers the historical tendency of the world system with its accumulation strategy and outcomes that often leads to global polarization of classes and conflicts. Especially during the current era of globalizing tendencies, several structural processes emerge, and global migration is one such world-historical process whereby the crisis of world accumulation generates certain tendencies and outcomes. One of these outcomes that engenders global migratory movements is the environmental degradation that occurs that is the natural fallout of the global accumulation process. As we have argued in the previous pages we witness migration as a result. With advances in transportation and communication, global migration is facilitated and thus many social worlds are in motion.

The above structural outcome has to be coupled with the cultural transformations taking place as a result of this *conjoncture's* globalizing process whereby cultural worlds are shifting. In an era of global crisis, we often find growing cultural fragmentation leading to both indigenization and cosmopolitanism (see Friedman 2004). With very different shares of the outcomes of accumulation process and with transnationalization and deterritorialization taking place, those at the bottom of the vertical ladder of each national space who have not benefited from this globalization process increasingly feel disenfranchised and separated from those at the top who have been handed global rewards. Instead of being elevated up the vertical economic ladder, they experienced a further incursion of their political, economic, and cultural spaces by those *in their view* who are not indigenous (legal and illegal migrants) to their political-territorial space—hence, the growing resentment to the guest worker or the migrant. The socioeconomic and political policies of the various governments of these receiving countries have not helped either. They have added to the social distance whereby there is the emergence of enclaves or ethnification of the sociogeographic space generated by governmental policies. Again, this further widens the distance between these disenfranchised groups. Attempts at assimilation policies or multiculturalism by a specific nation-state does not help either in achieving cultural security and social stability, for these policies pitted against the homogenizing tendencies of globalization further fragment an already discordant social and cultural world.

Given the above, destabilization of these polities, and hence the world system, will continue and exacerbate further with the increase in environmental refugees expected in this new millennium in view of anticipated global warming and environmental changes that will occur. The globalizing world of the twenty-first millennium thus has its own set of destabilizers. In this case, from the growing intensity of the Nature-Culture relations to meet the accumulation dynamics of a globalizing world, we find the addition of global migration as one added dimension that will pressure the stability of the social system in the twenty-first century. As a world-historical process, migration thus appears again as it did during the fifth century as a transformative process.

Urban Shrinkage

One of the significant features of prior Dark Ages is the sign of deurbanization or urban shrinkage as a consequence of the slowdown of economic productive activities and trade disruptions and also population growth declines. On a world-historical basis in *The Recurring Dark Ages* this was shown to be the case in prior Dark Ages. Do we see signs of this deconcentration during the current era? Where does the urban population move to? This deurbanization process is rather complex in which one can find a definite loss in the population of an urban area, such as a city, due to the changing economic structure whereby dated production processes and commodity production in old industrial structures give way to newer production processes and new types of commodities. Furthermore, with relocation of production to cheaper sites with lower labor costs or fewer environmental regulations, urban decline in older cities is facilitated. Such a typical process occurred, for example, in the northeastern states of the United States and in parts of Great Britain (Couch et al. 2005; Ostwalt and Rieniets 2006). Other factors identified with urban shrinkage are reduced or negative population growth in the urban areas, wars, political disruptions, natural disasters, environmental disasters, and disease epidemics (Ostwalt and Rieniets 2006).

How then can I understand this process of urban shrinkage that is coupled with the fact that there is ongoing urbanization occurring? At the turn of the nineteenth century, it has been estimated that only 2 percent of the world's population lived in urban areas. By 2000, the urban population in percentage terms has reached under 50 percent. It is estimated by the year 2050 about 75 percent of the world population will be living in urban areas but *this growth is projected to end between 2070–2100 whereby the process of urbanization is expected to end* (United Nations Fund for Population Activities 2007). While this is going on, we also have urban centers shrinking in population size. In the 1990s almost a quarter of the world's large cities shrank (Ostwalt and Rieniets 2006). To be more precise, between the start of the twentieth century (1900) to the end of the millennium, a total of 322 cities in both core and periphery shrank in terms of population losses (see figure 3.6). It seems that shrinkage has occurred almost twice as much for the core zone in comparison to the peripheral zone (see figure 3.7). Total population losses for all cities showing shrinkage came to 35.205 million persons. Specifically for core cities, the population losses were almost three times as much in comparison to peripheral cities (see figure 3.8). Average losses for core cities over this time span was 127,000 while for the periphery it was about 78,000 persons.

Urban shrinkage is a historical process and calculations for this period reveal that shrinkage, for the core, has an average rate of duration of about twenty-eight years, and for the periphery it was about twelve years (see figure 3.9). It seems that there are certain phases whereby shrinkage started over the course of the last millennium (see figure 3.10). For the core, a phase of increase in cities showing signs of shrinkage started from the 1950s, reaching a peak between 1970 and 1980, and another phase starting again from the 1980s to 1990. For the periphery, there was only one phase starting from 1980 to 1990. Explanations to account for these

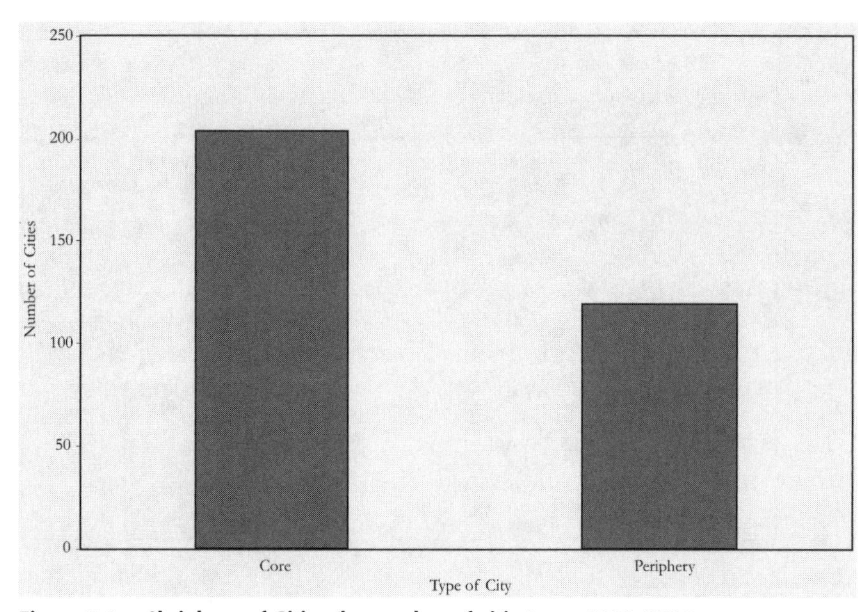

Figure 3.6. Shrinkage of Cities (by number of cities) A.D. 1900–2000

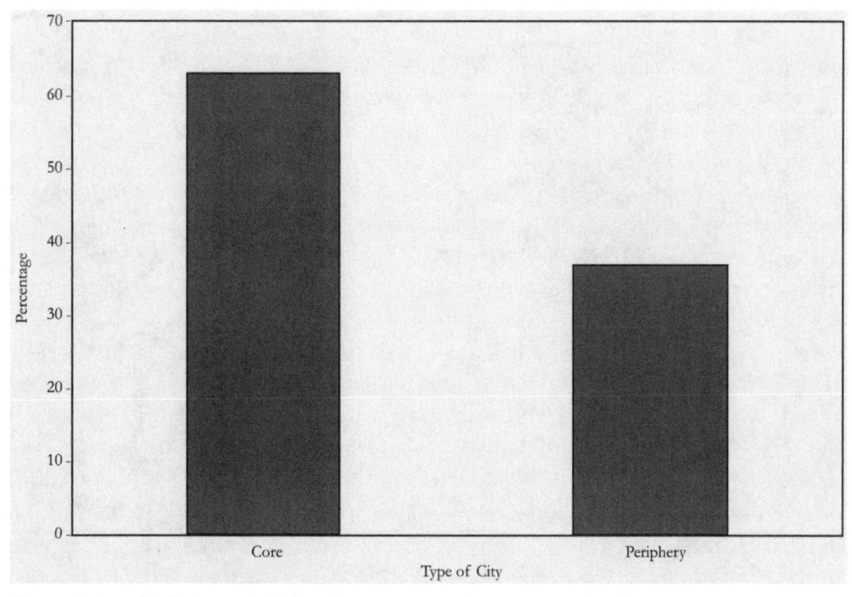

Figure 3.7. Shrinkage of Cities (by percentage) A.D. 1900–2000

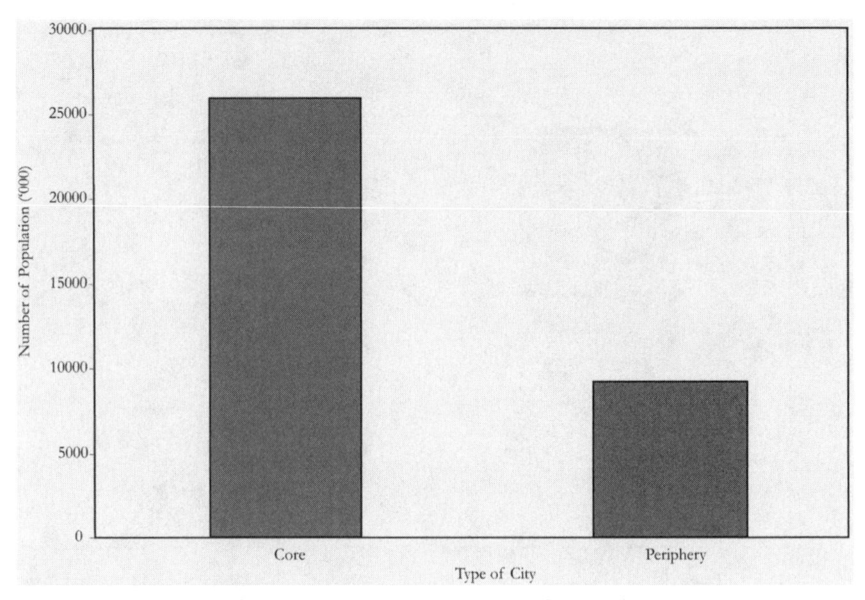

Figure 3.8. City Population Losses A.D. 1900–2000 (Thousands)

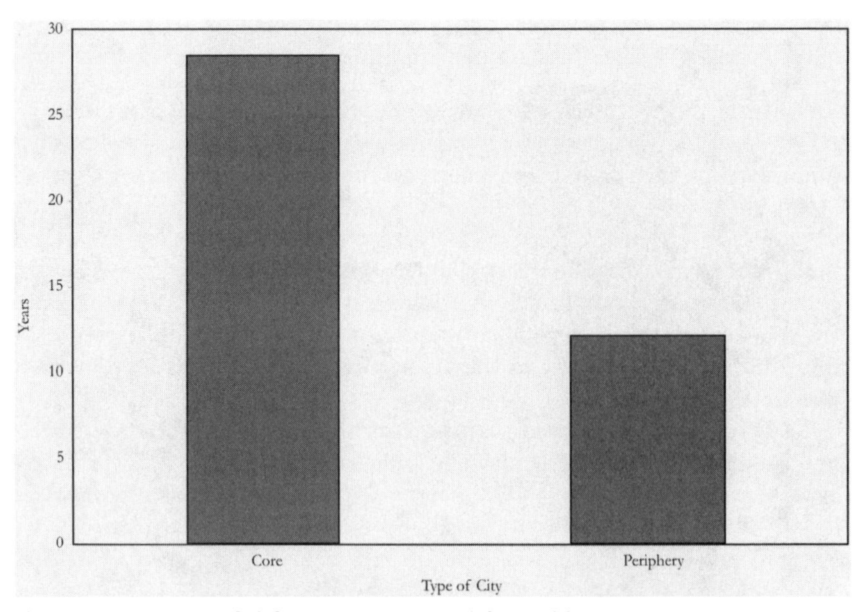

Figure 3.9. Average Shrinkage Years Core/Periphery Cities A.D. 1900–2000

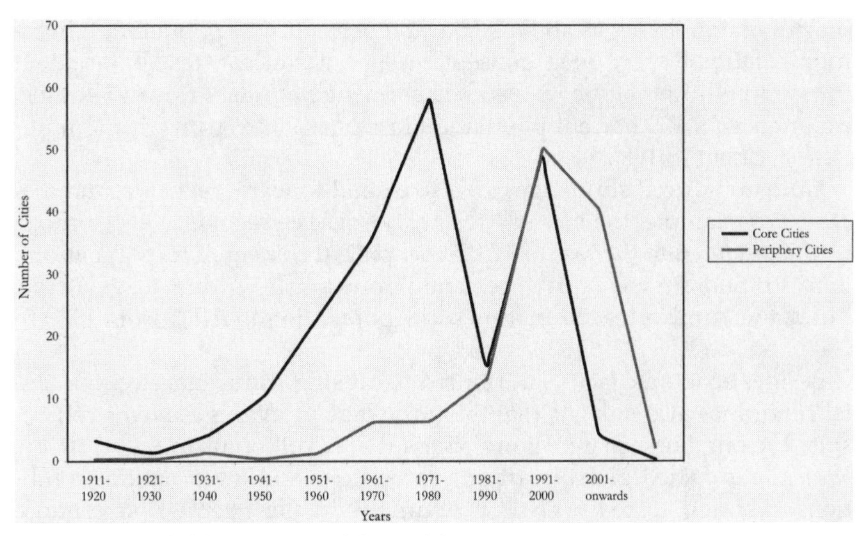

Figure 3.10. Shrinkage Core/Periphery Cities A.D. 1900–2005

trends can be as varied as economic downturns (a phase did occur in the oil shocks of the early 1970s, debt crisis in the 1980s for the periphery), wars, natural disasters, political dislocations, lowered or negative population growth, and environmental conditions. The works of Ostwalt and Rieniets (2006) and Couch et al. (2005) have provided some case studies using all of urban-shrinkage factors to account for some specific city declines.

For example, a comparison between Liverpool, England, and Leipzig-Halle, Germany, in terms of city shrinkage revealed major causes of a different nature. Liverpool experienced an economic downturn and the fate of capitalist economic accumulation, which meant a perpetual transformation of the industrial structure. Leipzig-Halle, which, since the end of the World War II, had developed an economic structure oriented around central planning and thus was sheltered from the vagaries of market forces, saw its shrinkage as a consequence of German political reunification that led to migration to the West where the rate of unemployment was lower.

Certain cities experienced shrinkage as a result of natural disturbances and environmental conditions. The Italian city of Messina never recovered from its earthquake of 1907. The city of Bam in Iran had a population loss of 41,000 (41 percent of its population) during the earthquake of 2003. More recently, the city of New Orleans lost 350,000 (72 percent of its population) persons as a result of Hurricane Katrina in 2005. The loss of ground water caused shrinkage in the city of Aralsk, Kazakhstan, where the population loss was about 20,000 (50 percent of its population). Environmental pollution as a consequence of a nuclear reactor accident through polluting effluents also has shrinkage effects. Pripyat, Ukraine, experienced a 100 percent population loss of 49,360 persons due to a nuclear accident in 1986.

Human-induced shrinkage can also extend to wars, genocide, and disease. For example, the city of Mostar (Bosnia-Herzegovina) experienced conflict that led to the loss of 12,000 people (10 percent of its population). Labado, Sudan, saw a 100 percent population loss due to genocide. Lusaka in Zambia lost 35 percent of its population to AIDS from 1980 to 2002.

Besides economic factors attributed to causing shrinkage, environmental conditions also indicate that this shrinkage process is not over yet, despite the ongoing urbanization process that is still occurring worldwide. With the expected global-warming projections and other natural conditions expected to reach crisis proportions in the twenty-first century, shrinkage will not only continue but also increase. Therefore, what we are providing in terms of preliminary data might be only the tip of the iceberg of the shrinkage phenomenon that has been going on in the last century. The United Nations Population Fund (UNFPA 2007) has already reported that urban growth in the latter half of the last millennium has consistently declined. If what has occurred historically in terms of deurbanization, the extent and pace of the process has yet to reveal itself fully to us, and the consequences of this process, as world history has shown, is not only transformative but also catastrophic in terms of collapse of societies and civilizations.

NOTES

1. See, for example, Wallerstein (1998) for his world-systems approach to such a transition.
2. See the debate between Theodor Adorno and Karl Popper on the logic and role of science and social sciences in understanding or explaining social life (Adorno et al. 1976).
3. Following through on this positivistic agenda, Lomborg used statistics (selecting time-slices, etc.) to pummel Brown's assertions or to show that Brown's assertions have no statistical backup, and therefore are "untrue."
4. This claim by the commission is based on the United Nations Food and Agricultural Organization's studies of the state of the world's forests, which Lomborg also uses for his deforestation statistical analysis and presentations.
5. The choice of the twentieth century as a break point is not based on any trends and tendencies for I have argued in *The Recurring Dark Ages* that Dark Ages are phases that are not cyclical in nature, and therefore it is difficult to predict the

start of the next Dark Age that will follow the Dark Age of Antiquity. On this basis, the twentieth century is chosen so that I can initiate a discussion on transition in view of the varied signals of natural system limits that have been identified by various studies, and the signs of global warming indicating climate changes that normally also accompany Dark Ages. A different periodization for a transition starting around 1967–1973 has been offered by Hopkins and Wallerstein (1996).

6. In certain ways, the provision of yearly endowments by Rome to the barbarian tribes to secure peace and cooperation in terms of troop provisions to Rome in times of war can be considered as foreign aid offered by the United States to U.S. "dependencies" during our contemporary period.

7. The hinterland incursions seem also to occur in higher frequency during other periods of Dark Ages such as during the crisis of the Bronze Age (Modelski and Thompson 1999, 261).

8. Heather (1991) has reported that the Goths would at any one time contribute about three thousand men to join the Roman military.

4

The Transformations

INTRODUCTION

System transformations are rare in world history. In *The Recurring Dark Ages*, periods of social and ecological distress the system traversed in its evolutionary trajectory were identified. Following such transformative periods, certain socioeconomic and political configurations emerged that structure the dynamics and processes of the social system. This assumes on the whole that ecological recovery to the devastated landscapes has occurred, and additional natural environments for the reproduction of social systems free of human exploitation exist in the postcrisis period. Regardless of the trajectories that arise from the reconfigured social system's dynamics and processes, their paths are still circumscribed by the *state and limits* of the natural environment, though they now operate within the regulative principles and tendencies of the reconfigured socioeconomic and political frameworks.[1]

Given that world history has exhibited these long historical moments of system transformation, can we again "let history be the teacher of life" (Schäfer 2007) and help us to understand and project possible ecological futures? Which historical pasts do we use as a point of reference? As stated in *The Recurring Dark Ages*, there were at least three long historical phases (2200 B.C.–1700 B.C., 1200 B.C.–700 B.C., A.D. 300/400–A.D. 900) of system crises. Structurally, one phase—the crisis of the late Bronze Age (1200 B.C.–700 B.C.)—transitioned from a system using bronze as a base metal to that of a system using iron as a base metal. This led to several structural changes that significantly pushed the social system along a set of socioeconomic and political trajectories that gave rise to significantly different pathways of social evolution and adaptation. In certain ways, the conditions and structural socioeconomic and political relations following the end of the late Bronze Age bear similar patterns and circumstances to our current times.

Let me restate the structural changes that occurred following the crisis of the late Bronze Age so they can be used as historical parallels to examine the historical circumstances that social and natural systems are currently experiencing (Chew 2006c, 169–90). They are the following:

- The replacement of base material from bronze to iron opened up economic opportunities and led to political instability as well.
- The use of iron led to increasing agricultural productivity (e.g., the iron plough), thus transforming social relations.
- A shift in economic and political structures from command-centered palace economies to mercantile systems occurred.
- "Polis" development led to new political structures in the Eastern Mediterranean.

If the historical specificities of the above four structural changes are striped away and translated into more generalized characteristic patterns, parallel transformations in the current socioeconomic and political conditions can be seen. It will also allow me to not only deconstruct the present occurrences and circumstances of its historical specificities (and thus to be able to differentiate between quantitative changes versus qualitative changes) but I also will be able to project possible futures or trajectories of the evolution of the world (social) system.

For the first structural change whereby the base metal of bronze was replaced by iron, one can generalize this shift to mean the condition of a scarce basic natural resource leads to its replacement by another base material, which will also lead to room for technological innovations and competition. Furthermore, the replacement could also mean that the original base material no longer has the same level of functionality as the replacement and hence the adoption of the replacement material. The new technological innovations provide opportunities for others to participate and hence the end of monopolization that the old technologies possessed.

The second structural change that emerged from the end of the Bronze Age saw an increase in productivity following the use of iron, which replaced bronze. Stripped of historical context, the adoption of iron can be viewed as a replacement of base materials and old technology which have led to a point whereby economic expansion can no longer occur as the existing technology and natural resource scarcity cannot generate further socioeconomic expansion. What then is suggested is the adoption of new base materials (iron) and social-technological innovations that open up new vistas for economic expansion and social organizations.

The third structural change when decontextualized suggests that the usage of new base materials and their associated technologies led to different forms of socioeconomic-political relationships as the opportunities for different levels and forms of participation emerge as a consequence of the resultant social-technological innovations.

The final structural change is not as connected in a structural relation to the others. Rather, as we have stated in *The Recurring Dark Ages*, it is an outcome of Dark Age conditions whereby the deurbanization process and the population losses that resulted provided the structural conditions whereby political ideas and participation could arise in small, less populous urban environments. This structural change can be realized as a consequence as small, less hierarchical isolated communities emerged from Dark Age conditions, whereas in the past—prior to the Dark Ages with palace command economies and kingships along with higher population—such a political arrangement would not have been possible. Decontextualized, the development of a different political arrangement suggests a search for political alternatives as a consequence of social-evolutionary development, conditions of scarcities, and political upheavals.

STRUCTURAL CHANGE I: NATURAL RESOURCE SCARCITIES, SILICON, AND INFORMATION TECHNOLOGIES

The global scenario of natural resource scarcities is increasingly becoming a major issue for the further expansion of the social system. This was discussed in the previous chapter in relation to various major studies from global to the national levels that have been undertaken to assess global supply. One of these impending scarcities is oil. Global oil production is supposed to peak in the first decade of the twenty-first century and is supposed to last from fifty to eighty years thereafter (Chew 2007; Duncan 2001; Deffeyes 2001; Campbell 2004). With the increasing expected economic growth in China and India, this decreasing resource will be a major issue of contention and competition for access, hence a volatile political-economic competitive environment. Already we increasingly find Chinese companies searching for and signing fossil-fuel contracts in Africa, Central Asia, Canada, wherever there is an oil resource that has not been locked by multinational petroleum companies from the core industrialized countries. With a world economy's socioeconomic reproductive processes highly dependent on petroleum, this impending scarcity will pose a challenge to find a replacement energy source before the supply of oil runs out. Such is the adaptive challenge that the social system faces.

Without an alternate energy source that can sustain the world's economy, the socioeconomic evolution of expansion will be curtailed.

Alternative energy sources from wind, solar, nuclear, biodiesel, ethanol, to hydrogen have been suggested and the transition to these replacement fuels will mean political intraclass conflicts (e.g., see Rifkin 2002). Ultimately, the final outcome will be decided by the eventual depletion of available oil supplies and the strength of the class fraction and/or perhaps regions that control the world's alternate energy sources and their associated technologies. At this juncture, a range of possibilities is there (e.g., see Hawken 1999). The outcome will depend on the socioeconomic and political scenarios following the collapse as I have envisioned in *The Recurring Dark Ages*.

Besides the natural resource scarcity issue, the functional capacity limits of certain base materials to meet the contingencies of socioeconomic needs should also be explored. The replacement of bronze by iron spurred technological innovation and production of tools and implements that facilitated increased production, for example, in agriculture for the post–Bronze Age period during the time we now classify as the Iron Age. The discovery and fabrication of certain composites with good resistance to high temperatures, oxidation, and brittleness have increasingly replaced standard aluminum, stainless steel, and iron in the manufacture of important components of machines and other items basic to our socioeconomic activities. Carbon composites, oriented polymers, fiberglass, and so forth have increasingly been introduced to replace aluminum and steel, which have in certain ways reached the limit of their functional capacities to enhance productivity and/or efficiency (e.g., see Sass 1998). Though costing more, these composites are lighter in weight. Thus, they have the advantage of being incorporated into products (such as airplanes, e.g., the Boeing 787 Dreamliner) whereby fuel costs and/or performance are important factors of consideration in the operation of the equipment or machine. On this basis, the opportunity for increasing efficiency and productivity accelerates.

The replacement for aluminum and steel with composites by no means is as transformative of the overall socioeconomic production process as the introduction of silicon in the fabrication of the microprocessor that now is the heart of the information technologies, thus having replaced the vacuum tube. Starting with the discovery of the transistor in 1947, the integrated circuit in 1957, the planar process in 1959, and the microprocessor in 1971, the pace of discovery in the information technologies has been of lightning speed. The use of silicon for fabricating a semiconductor for the manufacture of the microprocessor because of silicon's unique electrical properties has revolutionized social and economic activities of the social system. Silicon as a substance becomes the brains and memories of our computers.

The use of silicon leading to increased efficiency resulting in enhanced production levels parallels the introduction of iron during the late Bronze Age and early Iron Age as the base material when scarcity of bronze led to the adoption of iron for use in economic activities. For example, the use of the heavy iron plough provided the opportunity to work heavy clay soil that in the past was inaccessible due to the previous technology, thus leading to increased agricultural production. A higher agricultural surplus and a more efficient use of available land areas for production spurred the economies of the early Iron Age.

I see this also in the utilization of silicon in the manufacture of the microprocessor—through the use of the planar process in manufacturing it—which enabled the microprocessor to be used as the computer's central processing unit. This allowed the simultaneous storage of facts and information and the instructions or program on how to manipulate them. Not only can the microprocessor be used in computers, it can be installed in other apparatus that will enable the equipment to perform specific and varied functions, thus increasing the productivity and efficiency of the assigned tasks.

This technological innovation (transistor and microprocessor) replacing the vacuum tube enabled the introduction of modern-day computers. It was the beginning of what everyone has dubbed the Information Age (see Castells 1989, 1996; Clark 1997). This new age's infrastructure is anchored by computers and microcomputers over a backbone of communication lines of optical fibers replacing copper wiring. For the latter, the capacity to conduct digital data across long distances over very high speeds with very high capacities for information transfer has surpassed copper wiring in terms of the latter's functionality. For example, the copper wire on an integrated services digital network can carry at estimated 144,000 bits of information while an integrated broadband network built on optic fibers could carry a quadrillion bits of information. The earlier transatlantic cable line in use was only able to carry 50 compressed voice circuits whereas now with optical fibers it can handle 85,000 compressed voice circuits. This transmission-communication infrastructure is made possible also by advances in integrated circuit technologies and the digital switch pioneered by Bell Labs.

Parts of this communication infrastructure are also the communication satellites that act as anchoring points for communication networks. The Communications Satellite Corporation (COMSAT), which was introduced in the mid-1970s, enhanced communication between land and sea traffic, thus facilitating the shipping of goods. The International Telecommunications Satellite Consortium's (INTELSAT) IV-A carried phone circuits and later versions also had media communication and broadcasting. Later satellites enabled credit card verification and charging, further heightening the speed of credit charging globally.

With the fabrication of the microprocessor, it enabled Alan Kay of the Xerox Corporation to design the first microcomputer though it was never produced for the market. Mass market sales were only spurred by the availability of the Apple I computer offered by Steve Jobs and Steve Wozniak in 1976. International Business Machines (IBM) followed with its own personal computer in 1981. These two introductions to the marketplace led to a multiplicity of companies offering the personal computer for home and business adoptions. The software programs that accompanied the use of these machines were extensive. From personal home needs to financial spreadsheets to manage office functions and tasks rapidly transformed both the home and the workplace.

The advent of the micro processor led to major advances in computing power at the personal computer and microcomputer level that in time surpassed mainframe computers at a fraction of operating investment and costs. For example, the IBM mainframe 360/370 introduced in 1964 was surpassed in computing power by later versions of the microcomputer introduced in the early 1970s. As a result, not only were microcomputers used as stand-alone systems, they were networked for interactive communication and power sharing. The fact that the IBM personal computer had few proprietary technologies enabled the manufacture and proliferation of clone-type personal computers all over the world. As a result, there occurred the rise of major microcomputer companies in the United States, Europe, and Asia. Just like the use and manufacture of iron during the late Bronze and early Iron Ages, which occurred widely outside the palace-centered control of bronze manufacturing, we see similar conditions existing for the manufacture of the microcomputer.

The creation of the Internet culminates the network concept whereby the communication network is independent of command and control centers insofar as information and messages can find their own routes along the network and get reconstructed at any point of the network. Such an open architecture allows any individual, firm, or agency to participate according to their functions and intentions. It has led Castells to assert that "in 1999 there was no indisputable, clear authority over the Internet, either in the U.S. or in the world—a sign of the free-wheeling characteristics of the new medium, both in technological and cultural terms" (2000, 46).

With the formation of the Internet along with the advances in microcomputing and telecommunications as we have briefly discussed above, there was a shift to a system of interconnected information-processing devices. The computing power of this technological network is distributed in a communicative infrastructure supported by Web servers using common protocols of the Internet with access to other computers that serve as database and application servers. Access to this World Wide Web is no longer restricted to the microcomputer linked to a local area network, but is now open to other specialized communicative devices such as the cell or mobile phone. Such a distributive system of information with worldwide access via a number of communicative devices means that it can be used for a variety of functions and needs from personal messaging through to what is known as e-business and manufacturing. Furthermore, it can be accessed everywhere and anytime. This means that space and time no longer pose a great barrier to socioeconomic and political activities (Harvey 1990).

STRUCTURAL CHANGE II: TECHNOLOGICAL AND SOCIOECONOMIC INNOVATIONS AND OPPORTUNITIES

It is commonly agreed that the 1970s experienced socioeconomic crises characterized by falling profit rates of businesses, stagflation conditions, oil-price shocks, and high unemployment in core countries. Such conditions required the imposition of austerity measures at the national and global levels, including the use of monetary policy to control price inflation—the famous Volcker shock—that the United States was experiencing then.[2] To extricate the world economy out of a recession thus required policy shifts, reorganization of the production process, realigning labor relations, curbing inflation, and resuscitating consumer demand.

Clearly from a capital accumulation point of view, this required that capital appropriate a higher share and level of the surplus from production. This can be achieved through several ways. Attaining higher productivity through technological innovations is one route while at the same time policy changes that will lower taxation schedules could also be tried—the trickle-down theory. Profit rates can also be increased through decentralizing production to lower wage countries and regions and reducing further labor costs via rollbacks in social benefits and so forth. The latter can be undertaken both at the levels of the firm and national economies through changes in social-welfare policies and programs. State policies to support this shift can also be tailored to promote deregulation of the marketplace, regressive tax reforms, state support to high technology research and development (R&D), as well to defense and defense-related industries. The latter most often leads to technological discoveries that can be adopted for private-sector development and marketing. Furthermore, to further enhance opportunities for capital, the state undertakes the acceleration of the globalization process through economic-market liberalization and promotion. The latter entails the push to remove tariff barriers in the Third World for foreign investment and sales. This globalizing process enables global capital accumulation to penetrate and accumulate without any restraints and provides the opportunities for twenty-four-hour global investment opportunities that result in a more rapid turnover of capital. New markets are formed and different types of verticalization of production can be pursued.

All of the above general strategies were tried and, in most cases, implemented within the broader global strategy of regionalization and globalization of economies and polities. Rather than covering this broad range of strategies and themes as described above, for this section on structural change focusing on technical innovations leading to the expansion of the economic sphere, we concentrate on conditions and circumstances following the introduction of silicon and the subsequent rise of information technologies that provided the capacity and opportunity during a conjuncture of globalizing forces and a world economy in recession.

If we gaze on the various studies on the emergence and growth of information technologies, there is the acknowledgement on how these technologies have increased the productivity in the workplace and the production process. The more significant argument that we need to pursue is that of the emergence of a new economy based on informationalism that is structured via networks (Castells 1989, 2000; Aneesh 2006). What this means is that the social structure and social fabric of society are restructured to the extent of the displacement or compression of the spatial and temporal dimensions we have been used to. Transformations in the workplace occur as a result. Call centers located in parts of the periphery respond to customer queries when the work day ends in the core countries. This has led to customer service being available twenty-four hours a day, taking advantage of the different time zones. Like financial trading now operating on a round-the-clock basis, the innovations in information technology have enabled corporations to take advantage not only of the differential in labor costs but also the differences in time across the world economy.

By no means do the innovation and transformation of the workplace stop here with labor in call centers that is most often quite low skilled. The availability of highly trained labor at low cost has also led to the transfer of skilled work processes to other parts of the world economy, thereby lowering the costs of doing business. Aneesh (2006) has defined this decomposition and recomposition of the labor process as one whereby labor does not move as it did in the past. Instead, labor stays put and it is the work that has shifted over the information network. We have, in Aneesh's terms, virtual migration instead of labor migration: "Using high-speed data communication links, programmers based in their national territories can work on line and in real time on computers situated anywhere in the world, thus obviating the process of material migration for both labor and corporations" (2006, 35). In sum, production processes, lifestyles, and socioeconomic activities are being transformed globally in line with various properties of a network society.

Based on such a structure of an information society organized along networks of activities, productivity becomes increasingly knowledge based, and knowledge and information themselves become commodities that can generate economic activity, surplus, and profits (e.g., see Hardt and Negri 2000). What this means is that symbols and information codes produced by knowledge become available as "matter"—though different from physical matter (Nature) but has as much a material force in terms of its capacity to generate value and surplus in a technological society (see Baudrillard 1975, 1983).

Productivity becomes an issue of the capacity of economic units to generate, create, process, and apply knowledge-based information. Such a model of the new economy opens up new vistas for diverse economic activities to be undertaken, and for creative minds to come up with as many processes, pathways, and products (material or information) for exchange, sales, and so on. We can see the expansion of various possibilities in view of the existence of the World Wide Web with its open architecture void of centralized control and increasingly the call for the development of open source codes for the programs that power and manage this technological universe. Such emergence of an electronic network along with the nature of information technologies whereby its production and management are undertaken on a global basis and in a decentralized network thus generate the conditions for various economic opportunities for different communities, economic units, or agents to be able to participate, to some extent, in this economic exchange. One can see this as the widening of the participation of the economic marketplace that lesser groups or economic agents can enter just as long as they have some technical expertise where knowledge and innovative technical abilities become more the key to success than economic strength or power, for example, Indian high-technology companies (e.g., see Aneesh 2006; *International Herald Tribune*, September 24, 2007, 3). In this sense, the possibility of economic success can now go beyond the core countries and their transnationals that historically have monopolized the economic marketplace. Of course, after having stated this, it does not mean that this economic space is totally open. Certain conditions and factors, such as technical knowledge and capacity, synergistic relations between knowledge-producing institutions and economic agents or groups, supportive state policies, and availability of start-up capital, are still required for a successful economic participatory outcome. This means that there is an opening up of opportunity for economic units and agents beyond advanced industrial countries, but it does not mean the marketplace is totally open without any required conditions for success.[3]

On this basis, silicon as a base material coupled with the microprocessor, besides providing the capacity to greatly improving efficiency and productivity in the usual commodity production processes, have also opened up new areas, markets, and avenues for economic expansion that the old technologies and socio-organizational frameworks could not provide. In this context, this parallels in certain ways the adoption of iron manufacturing at the end of the Bronze Age that enabled the fabrication of the iron plough, which led to economic expansion by opening up regions whose soil conditions were not practical for agriculture in the past.

THE TECHNOLOGICAL NETWORK

With the arrival of information technologies as the underlining basis whereby socioeconomic and political activities are framed, new horizons in terms of economic opportunities open up that did not exist in the previous economic marketplace, such as fabricating the electronic machines and media that allow us to communicate and store information within the electronic space, services providing support to such activities, and the various functional services to analyze, modify, and compute the information including the technical development of software programs to run the information technology system and networks. Besides this, the various strategies that generate further economic opportunities through the innovative adaptation of the functionalities offered by the technological electronic framework, such as e-business on the Internet, need also to be realized.

The Hardware

Even though the introduction of the computer as a data analysis device was available in the mid-twentieth century, the proliferation of data analysis and storage devices did not gain widespread adoption until the commercial introduction of the affordable microcomputer for personal use. From then on, a vast array of equipment for various technological functions became available for purchase. If one examines the type of possible electronic equipment that can connect onto an electronic network, the following are possibilities: microcomputer, terminal, scanner, printer, modem, phone, wireless PDA, satellite dish, video camera, PDA cradle, servers, and mainframe—and by no means are these electronic gadgets the only equipment used.

Given this array of possibilities, the economic opportunities to produce the equipment for sale is extensive, especially when global production and sourcing further enhance the profit generation. As we have stated in an earlier section—Structural Change I—the advent of the microprocessor has also led to the adoption of it to almost every piece of equipment that we use in our everyday life, from the stove, washing machine, automobiles to the tooth brush. This means that a set of new ancillary industries emerge that feed into the global production chain.

The new array of electronic equipment that enables us to function and communicate with others in work and personal situations comes accompanied with a requisite vast array of extension and adaptor-type equipment from the surge protector power bar, to various types of wiring for handling, for example, Ethernet connections, wireless electronic speakers, mouse pads, monitor screen protectors, and the list goes on. Again, this is a series of gadgets and items that would never have been available in the marketplace if not for the technological network. In addition to these ancillaries, a range of equipment is needed to form the architecture of the network that has their associated product types, from digital routers to fiber optic cables. Of course to make these gadgets and machines, other machines are needed.

There are various types of media to store this new information. A short listing exhibits the array of data storage options: cassette tape, CD, DVD, flash memory, floppy disk, hard disk, microdrive, worm drive, zip disk, etc.

The Software

Ever since the success of the Microsoft Corporation with its software providing the operating system of the microcomputer, the range of software available to manage, coordinate, store, and analyze data has grown geometrically. In this respect, the use of knowledge and information to generate a knowledge-based product for market exchange depicts the uniqueness of this new economy. In a broad sense, knowledge is exchanged for money in the marketplace. Knowledge becomes a commodity for sale, and in most cases today, there is not even a "material" product that is exchanged other than symbols and codes of a software program downloaded from the World Wide Web for a monetary amount transferred electronically instead of money in its paper or metallic form. This is distinct from the provision of services where one is charged a monetary amount, such as banking, for example. The latter has been going on for thousands of years with the emergence of the market.

Besides the range and types of program software available for operating our individual electronic devices and for access to the Internet, there are other types of software that have been created to write other software to meet some system technological functions and also to build, design, and maintain portals and Web sites. One can see that there are no limits in terms of the expansion of this software area and, with the proliferation of e-business, the range of software needs is expected to expand.

Data, Information, and Services

This sphere of the technological framework has increasingly expanded and, with the capabilities of what information technologies can provide, the intensive and extensive expansion of this sphere has no bounds. Data collection has been a human activity since the early civilizations whereby copious records of palace or temple trade transactions were kept, in ancient Mesopotamia for example. Data collection has been enhanced by the capacity of storage media and the ease of storage. Along with the growth of bureaucratic functions in the public and private sectors spurred forward by the expansion and "success" of science and management techniques, data collection has been widespread, especially when everything is now easily recorded with bar-code scanning that has pervaded every product and service that we use every day.

If we consider the data lifecycle as a way to account for the range and type of services that can be provided or created, it gives us a rough idea of the extensiveness of economic opportunities (e.g., see Bergeron 2002). Capturing or creating data is the first step in data collection. Here, the recording of an image, sound, character, or symbol requires data collection functions or services. Following capture, the data is stored and in this context data storage facilities are available in terms of the level of sensitivity and importance of the data. Most often this is the archiving of the data collected. Raw data collected is modified or edited. If the data is textual, it is spellchecked, formatted, and, if needed, combined with other texts and images. Images and audio are processed to meet their application requirements. In some cases, the images need to be resized to meet certain use requirements—for example, images used for print media is of a different size than those needed for the Web. The colors of the images are also chosen carefully because they are affected by the viewing monitor's color configurations.

The part of the cycle that is the most important economically is the data modification phase. Depending on the functional need the data is exchanged or distributed. In an age of information where information is, in certain ways, both money and power, access to it provides the economic unit or agent the capacity to maximize the socioeconomic and political exchange. Digital data in this context becomes the currency. In the private sector, every point of contact between the customer and the business is critical in terms of the business's ability to provide a product or service, as well as to maintain customer loyalty. This would be the basic level of data (biographical records, financial data, tax records, etc.) used by a business entity or public agency. To this extent, the quality of the data and the stages it goes through in terms of cleaning and manipulation adds value to the data. Insomuch as the data is properly formatted already adds a value to it in terms of its quality when it is retrieved for use in different economic functions. When certain data sets are aggregated to form a base of potential purchases of commodities or services, for example a mailing list, its value goes up in dollar terms. The other type of value-added data that has economic value is what is termed *data in context*. Here data that has not been aggregated but is context specific is modified, for example for an end user, directions to a hospital or how to find a physical location. This type of data is increasingly provided for use in equipment and machines as our everyday activity becomes more and more automated, thus eliminating economic positions that would have otherwise provided this data flow in the past.

With such an extensive array of information possibilities in terms of data capture, storage, and use, data mining as economic or security activity is increasingly becoming a major focus undertaken by business entities or public agencies to meet their mission objectives, whether it is to enhance profits or improve service efficiency and productivity. These activities can cover a broad area of socioeconomic activities from customer preferences and habits to electoral preferences and philanthropic donations. In some cases, a customer's relationship with the private business in terms of how much profit the customer has provided to the company historically determines the type of service the customer receives from the business when phoning in. Those who have generated large profits have their phone calls answered immediately while those who have not are kept on hold.

E-business

With the World Wide Web, the electronic marketplace, which exists in the virtual world, becomes an economic reality. Two types of economic transactions usually occur: B2B (business to business) and B2C (business to customer). It is estimated that in 2004 global e-commerce totaled $6.8 trillion, of which 90 percent was B2B (Castells 2000). The way the business enterprise conducts its business sales and exchange is shaped by the nature of the technology and the technological network. The business model of Cisco Systems is one such example of a B2B operation. Cisco's business products cover the supply of the equipment that forms the backbone of the Internet. Its business operations involve its customers, its suppliers, its partners, and even its employees who are connected to each other via an extensive technological network. By having a network of suppliers online, Cisco does not really manufacture most of the equipment it sells. Rather its suppliers' manufacturing plants become part of the Cisco Systems' network of manufacturing. Cisco's Web site provides suppliers, customers, and its employees the capacity to order (if you are a customer), supply (if you are a supplier), and support (if you are an employee of Cisco providing service) Cisco's product lines. There are very few person-to-person interactions unless they are major contracts. Once an order is made, the order is filled online by Cisco's network of suppliers who are also connected to Cisco's networks. From such a setup it is clear the level of efficiency that can be achieved and the cost savings that can be obtained have accelerated globally. Furthermore, the amount of information collected providing opportunities to meet market demands and market slumps can be better anticipated. A business model using information technology and the Internet can also be found in computer manufacturers such as Dell or Hewlett-Packard. Other sectors of the businesses such as in agricultural machinery (John Deere), warehousing and construction (Bechtel Group), books (Amazon), auction (e-Bay), and transportation have also appeared. In fact, these days it is almost impossible to find a business entity in the world not having at least a minimal capacity to offer e-business via the Internet to its customers. The level of integration of business operations to the functionalities of the Internet is quite wide ranging.

A growth area of e-commerce is in the field of mass media. With the capacity to access and integrate various modes of communication and entertainment into an interactive framework has spurred remarkable innovations and experimentation. Television, film, music, and print are now widely available for downloading off the Internet and exchanged. Not only the availability of different mass media but also its availability has led to a global marketplace for this cultural space through satellites and real-time. The introduction of various electronic gadgets from the VCR and DVD to iPods means the opportunity for mass media communication and dissemination. Increasingly, the opportunity to download from the Internet due to technical advances means that information (such as electronic books) and images are commodities available for sale without any materiality being exchanged. Of course the accessibility to these mass media connections and opportunities is limited to those who can afford economically to gain access. It means that the economics of the digital divide still plays a part in this technologically framed world of mass media.

Clearly from the above we can see the wide array of economic opportunities and expansion that have opened with the advent of the microprocessor and the use of silicon.

STRUCTURAL CHANGE III: ECONOMIC OPPORTUNITIES IN AN INFORMATIONAL WORLD

If information and knowledge form the basis by which value is obtained in this informational economy does it mean that the control of the world economy continues to remain in the hands of the core zone for the future, or does it mean that there can be some shifts whereby others outside the core zone can find opportunities in the global marketplace? Clearly, there are some preconditions and socioeconomic and political processes that need to be in place in order to seize these opportunities.

The global digital divide already precludes some, especially in the periphery, to have such an opportunity (Castells 2000; Kagami et al. 2004; Wilson 2004; James 2003). If one examines the territorial distribution of Internet users on a global basis, there is a great degree of unevenness determined by the availability of technical capacity and increasingly being determined by access to large broadband networks. In year 2007, there were about 1.244 billion daily users (see table 4.1). This level of usage amounted to about 19 percent of the world's population. The overall increase in terms of Internet use has increased by 245 percent since year 2000. Asia is the most dominant region of the world that has access to the Internet with about 459 million users, followed by Europe with about 338 million users. North America ranks third with 235 million users. The peripheral areas make up the rest with Latin America (115 million), the Middle East (34 million), Africa (44 million), and Australia and Oceania (19 million) (www.internetworldstats.com). Usage has grown tremendously from year 2000. For some regions, growth has been astronomical as most of them came "late to the game." For example, between 2000 and 2007, Africa's usage grew 874 percent. The Middle East had growth of 920 percent, and Latin America and the Caribbean had a 540 percent increase. Asia had about 300 percent increase, whereas Europe had an increase of over 200 percent. North America had the lowest

Table 4.1. Distribution of World Internet Users, 2007

Region	*Percent World Population*	*Percent Population Penetration*	*Percent of World Usage*	*Usage Growth 2000–2007*
Africa	14.2	4.7	3.5	874.6
Asia	56.5	12.4	36.9	302.0
Europe	12.3	41.7	27.2	221.5
Middle East	2.9	17.3	2.7	920.2
North America	5.1	70.2	18.9	117.2
Latin America, Caribbean	8.5	20.8	9.3	540.7
Australia, Oceania	0.5	55.2	1.5	149.9
World	100	18.9	100	244.7

increase with just over 100 percent. Overall, total world usage increase was 244 percent.

World distribution of users for year 2007 was Asia with 37 percent share, Europe at 27 percent, followed by North America at 19 percent. Africa, Latin America-the Caribbean, Middle East, and Australia-Oceania make up the rest with 17 percent of global share. If we compare these figures to what Castells (2000) reported, Asia has overtaken North America's share—in year 2000 Asia's share was 20.6 percent whereas North America's share was 42.6 percent. Distribution of users on a worldwide basis does not reveal the percentage of users as a proportion of the total population of the country or region. Even though world-user share for North America is only at 19 percent, over 70 percent of the total population of North America uses the Internet. This makes the region with the highest penetration of Internet users. Australia and Oceania is second with 55 percent, and Europe is third with 42 percent of its population Internet users. This is followed by Latin America and the Caribbean with 21 percent and the Middle East with 17 percent of their population using the Internet. The lowest penetration of Internet users are Asia and Africa with 12 percent and 5 percent respectively, even though Asia has the world's largest number of Internet users because of its population size.

Access to the Internet is uneven within region and within a country with urban areas having a high level of access regardless of core or periphery. The top five countries in the world in terms of Internet users are the United States (211 million users), China (162 million users), Japan (86 million users), Germany (50 million users) and India (42 million users) (see table 4.2). In Asia, for example, penetration as a percentage of the country's population reflects the unevenness. The highest penetration in terms of number of users are located in Japan (68 percent), Hong Kong (68 percent), South Korea (67 percent), Singapore (66 percent), Taiwan (63 percent), Malaysia (53 percent), Brunei (41 percent), and Macao (40 percent), which are by no means the most populous countries in Asia, while India (5 percent) and China (12 percent) are lagging behind (see table 4.3).

Access does lead to control of Web sites' content on the Internet. The production of content suggests the capacity of country or region to dominate the Net and thus the opportunity to participate in the informational economy that underlines the global economy. A survey of content producers in 1999 found that the United States had 54.6 percent of the world's Internet domains, followed by Germany (6.8 percent), the UK (6.5 percent), Canada (4.7 percent), France (2.2 percent), the Netherlands (1.7 percent), Denmark (1.6 percent), Italy (1.6), Japan (1.5 percent), Switzerland (1.5 percent), Sweden (1.4 percent), Brazil (1.3 percent), Argentina (1.2 percent), Australia (1.1 percent), and others such as Spain, Austria, China, South Korea, and South Africa with less than 1 percent (Zook 2000,

Table 4.2. Internet Users (Broadband) by Top 20 Countries, 2007

Countries	*Number of Users (Millions)*
United States	211
China	162
Japan	86
Germany	50
India	42
Brazil	39
United Kingdom	38
Korea	34
France	33
Italy	31
Russia	28
Mexico	23
Canada	22
Spain	20
Indonesia	20
Vietnam	17
Turkey	16
Australia	15
Taiwan	14
Philippines	14

2001). Such dominance is being reduced in content production, though the dominance in design and content will stay for some time (Castells 2000).

The technical capacity and infrastructure (as characterized by Web site content production) by no means are the only preconditions that will enable a greater participation in the global marketplace by business entities outside the core zone of the world economy. There are other factors such as the availability of knowledge-producing institutions for supplying innovations as well as a large pool of scientific and technical labor, state support for R&D, a milieu of innovation, and availability of investor funds (venture capital) (Castells 1989). In this case, Silicon Valley in California is an example of such a milieu. The factors are linked in a synergistic fashion that leads to an environment of innovations and productivity. No doubt, such an identification of these factors would mean that they are not easily found everywhere in the world system, especially in the periphery of the world system. On this basis, if such are the necessary conditions to take advantage of the transformed world economy following the advent of the silicon and the microprocessor, then it looks like the hierarchical order will likely still remain. The World Economic Forum's

Table 4.3. Internet Usage in Asia by Country, 2007

Country	*Penetration (% Population)*	*Percent Users in Asia*	*Use Growth 2000–2007*
Afghanistan	2.0	0.1	53,400.0
Armenia	5.9	0.0	476.0
Azerbaijan	9.8	0.2	6,809.2
Bangladesh	0.3	0.1	350.0
Bhutan	3.7	0.0	5,900.0
Brunei	41.0	0.0	452.0
Cambodia	0.3	0.0	633.3
China	12.3	35.3	620.0
East Timor	0.1	0.0	0.0
Georgia	7.6	0.1	1,560.0
Hong Kong	68.2	1.1	113.7
India	5.3	13.1	1,000.0
Indonesia	8.9	4.4	900.0
Japan	68.0	19.1	85.9
Kazakhstan	8.5	0.3	681.4
North Korea	—	—	—
South Korea	66.5	7.4	79.2
Kyrgystan	5.5	0.1	477.7
Laos	0.4	0.0	316.7
Macao	40.1	0.0	235.0
Malaysia	52.7	3.2	302.8
Maldives	6.6	0.0	235.0
Mongolia	10.3	0.1	794.3
Myanmar	0.5	0.1	29,900.0
Nepal	1.0	0.1	398.8
Pakistan	7.2	2.6	8,861.9
Philippines	16.0	3.0	600.0
Singapore	66.3	0.5	101.8
Sri Lanka	2.2	0.1	252.3
Taiwan	63.0	3.2	131.6
Tajikstan	0.3	0.0	875.0
Thailand	12.6	1.8	268.1
Turkmenistan	0.9	0.0	3,140.0
Uzbekistan	6.6	0.4	23,166.7
Vietnam	20.3	3.7	8,510.0
Total Asia	12.4	100	302.3

2006–2007 *Global Information Technology Report* further confirms this hierarchical order. Of the top twenty countries in the world ranked for network readiness, there is no developing country though there are some newly industrialized countries (NICs) (Singapore, Hong Kong, Taiwan, and South Korea). The ranking of network readiness is based on the information and communication technology capacity of 122 nation-states ranked along the dimensions of environment (market environment, political and regulatory environment, infrastructure environment), readiness (individual readiness, business readiness, government readiness), and usage (individual usage, business usage, government usage). Without a doubt, with only a few exceptions (Malaysia, ranked 26; United Arab Emirates, ranked 29) there were no Third World countries in the top ranked thirty countries.

There are, however, some promising signs whereby a selected few (individuals, companies, countries) might have this opportunity. Take for example, the Internet whereby information can be turned into a commodity for sale and thus economic opportunities, the architecture of openness that has depicted the Internet since its inception and the call for open source code (as exemplified, for example, by Linux) the possibility exists for others (i.e., those individuals or companies outside the core) to participate in this increasingly informational economy in terms of products and services. Again, if the development of Linux is an example, clearly large capital investments are not as necessary if the condition of open source code continues. Once the innovation has been generated, there will come into play the adopters and adapters that will lead to knowing the kinds of products and processes that can be developed from such an innovation. In other words, open source networks can lead to benefit those who participated in the network of communication whether these individual or business entities are located in the core or periphery, though each has to have an adequate level of technical knowledge of information and expertise. Their successes will, of course, be conditioned by the factors (milieu of innovation, venture capital, etc.) we identified in the previous paragraph. Profits and market shares can be received from designing applications, manufacturing gadgets associated with the innovation, selling services, and customizing innovation for clients.

Finland is perhaps an example of a country that has benefited from the potentialities of the informational economy. The country as a whole is one of the most technologically advanced. Mobile-phone ownership is at the level of 80 percent of the total population. Its success is exemplified by its phone company, Nokia, and the open source operating system, Linux, which is a challenge to Microsoft's operating system. Finnish economy in the last five years of the twentieth century had a growth rate of 5.1 percent GDP. The economy was transformed from a largely resource (metals and wood) export economy in 1960 that was 70 percent concentrated in this area to a 30 percent concentration by 2000. Besides this, another distinctive feature of the Finnish economy is its welfare state whereby a generous social system of support (education, health, retirement, unemployment insurance) exists for its citizens. This Finnish social program has not been curtailed in a world whereby some of the advanced economies have retreated from this social initiative with the advent of neoliberalism as an economic agenda.

The information technology sector of the Finnish economy plays an important role in producing advanced technology (Castells and Himanen 2002). It accounts for 45 percent of the Finnish GDP. There are about three thousand informational technology companies in Finland, specializing in software and wireless, hardware, and network products for business applications and personal use. The success of the Finnish model rests in a synergistic mix between business innovation and openness with state support via policies and funding for research and development (Castells and Himanen 2002). For the latter, state-led policies developed innovative institutions, strengthened basic and applied research in the universities, allowed for the development of individual innovations, and encouraged a national identity of technology favoring instead of technology resistance. What this shows is that a welfare-activist state is not inimical to the development of an information-technology economy. By redistributing the social rewards, it ensured the foundation of human productivity through the provision of health insurance, education, and other life-enhancing qualities. Taxation thus is not seen as an economic problem provided the economy is booming. Through its policies, the state also provided mediation between labor and capital, ensuring work flexibility and stable industrial relations.

In view of the success of the Finnish business model in adopting informational-technology frameworks and the active role of the state in assisting to achieve these goals, Finland's success could be an example and also a hope that others might have the opportunity to follow this path in view of the nexus that has opened up with the advent of the Informational Age and network society. As Castells and Himanen have expressed:

> The Finnish experience also offers some hope for countries currently stagnating at a much lower level of development around the world. In contrast to the image of Finland as a rich, Scandinavian country, it must remembered that only three generations ago Finland was a very poor country, with most of its population in agriculture, largely dependent on its forest resources, only loosely integrated into the main channels of capital, markets, and technology in the world, and with a very limited public coverage of people's needs. It was, overwhelmingly, a poor agrarian society surviving in harsh climatic conditions. The ability to leapfrog in about half a century from the depths of economic backwardness to the forefront of informational development shows that it is not historical fate. (2002, 169)

Another route whereby individuals, companies, and nations can take advantage of the informational economy is represented by what Saxenian (2006) has dubbed "brain circulation." In this case, she is documenting the success of Taiwanese, Indian, and Israeli entrepreneurs in establishing informational companies in their respective countries of origin following a long sojourn studying in the United States for graduate degrees in information technology and engineering, followed with work in the informational technology sector in the United States, especially in Silicon Valley (California). These "new argonauts," often the best and the brightest who have earned their graduate degrees in science and engineering, most often ended up working for America's fast-growing technology companies (Saxenian 2006). Through social networking in America with their own ethnic colleagues who were also working in similar occupations and with their own counterparts in their home countries, the new argonauts have been able to establish companies in their respective countries of origin taking advantage of the economic openings now offered by their countries of origin. At the same time, they have been able to straddle and take advantage of the resources available in America from their past American work experiences and social networks. These opportunities include venture capital investments, access to legal and monetary investment banking, hiring managerial and technical talent from the United States, and raising public capital on U.S. stock exchanges. For their countries of origin, besides establishing successful technology companies that also have offices in the United States, the new argonauts have provided advisory capacities toward improving local infrastructures such as universities and research and training institutions, forums for information exchange among local technology companies, and advising their respective national governments on legal and market reforms so that these countries can be competitive on the world market.

Instead of a brain drain effect that was used to describe the loss to the Third World of foreign graduate students who never returned to their countries of origin following graduation, with the structural changes in the global economy that have occurred in the last few decades, we are witnessing a circulation of the knowledge received (Saxenian 2006). Besides those who have returned to establish information-technology companies, there are many more technologically trained professionals who have returned to Taiwan, China, India, and Israel following graduation and/or who have worked for a while in technology companies based in the United States because of the interesting and rewarding openings that are now increasingly available to them.

The previous paragraphs suggest that given such a conjuncture of structural opportunity when the informational economy is yet to mature fully and that *information and knowledge* generates value and economic expansion, what appears on the landscape is an opening of a nexus whereby innovative ideas can generate future economic success. On this basis, the economic "marketplace" becomes open to those with information and knowledge thus perhaps "democratizing" the business landscape to an extent when in the past the "old" industrial order reigns. Such a historical *conjoncture* is very much like that of the end of the late Bronze Age when the old palace-command economies were on the wane with the emergence of the use of iron and the rise of mercantile systems in the eastern Mediterranean.

STRUCTURAL CHANGE IV: POLITICAL RECONFIGURATION

Over the course of the late twentieth century, several political-economic projects were proposed that were different from that of the popular configuration of the nation-state that emerged with much earnest following the end of World War II. Whether these then new political projects were introduced due to a concern for political security and stability, environmental degradation, and the scarcity of natural resources, they offer new political orientations with regard to different configurations that the social system could organize. As a result, they offer different political choice of orientations and actions due to the nature and level of political organizations that have been proposed or implemented.

If one examines the various political configuration proposals that have been made, two very different suggestions emerged that we should examine more closely for they are likely political choices that would be offered in this new millennium depending on the contingent conditions in the social and natural systems. One that has been proposed, which in general terms can be categorized as "regionalization," suggests a political framework that transcends the political organizational level of nation-state that is the most common political configuration in current circumstances. The other, which is toward decentralization and can be categorized as "localization," is the exact opposite of the "regionalization" project for its configuration proposal is toward the opposite and very much a shift to a more localized level and very much a movement away even from the nation-state level of political setup.

These two political rubrics find expression in the last century in the European Union (EU) for the regionalization project, and bioregion for the localization project. Both of these two political projects have different historical developmental origins and different plans for their adoption for the future in other parts of the world where these ideas have not taken hold. On this basis, let us examine each historical-political practice in greater depth in order that we might anticipate the types of political configurations that will most likely occur in the twenty-first century and beyond, which we will be discussing in the final chapter of this book.

Regionalization

There have been numerous regionalization projects that have occurred since the end of World War II. This time breakpoint does not mean that there were no regionalization projects in world history. Indeed they were numerous if one surveys the course of world history, for example, the Athenian League (478 B.C.).[4] The time break is chosen primarily because we suspect that system transition is in the process of becoming more and more a historical possibility in light of the conditions in the natural and social systems.

Various regions of the world since World War II have embarked on a series of regionalization projects mostly to secure a common economic cooperative system via trade. For example, in West Africa one finds the Economic Commission of West African States (ECOWAS), in Southeast Asia there is the Association of Southeast Asian Nations (ASEAN), the North American Free Trade Association (NAFTA) in North America, in Latin America there is the Andean Pact, and finally in Europe, the European Union (EU). On the whole most of these regionalization projects are mainly oriented toward economic considerations, with the exception of the EU, though at the start, the EU was mainly focused on economic cooperation too.

On this basis, let us examine the EU project in greater detail to understand the dynamics of such a political configuration (Keohane and Hoffman 1991a, 1991b; Waever 1995; Ruggie 1993; Sachwald 1993; Bidelux and Taylor 1996; Hill 1996; Hirst and Thompson 1996; Habermas 2001, 2006). Started in 1948 with the meeting of European leaders in The Hague to discuss the prospects of European integration, the aim underlying the discussion was to avoid a future war. The causes of World War II were numerous, but one of the ones referred to in chapter 3 was Germany's search for additional natural resources and precious metals to meet its economic needs and political objectives. This cause ties the proposal for a different political configuration in 1948 to one that has its basis in natural resources. Having stated this, this early initiative went through a lot of changes as the idea of a different political configuration continues to form as a result of a number of events and circumstances that changed the socioeconomic and political atmospheres of Europe.

The initial move toward integration was the formation of a common market for coal and steel industries for European countries (France, Italy, West Germany, and the Benelux countries) involved in this political project in 1951. Following this was the Treaty of Rome in 1957 that led to a common policy on nuclear energy and the formation of the European Economic Community (EEC) for improving trade and investment among the member countries. Of course, such community building was conditioned by different political objectives of the different heads of state. President Charles de Gaulle of France added his political objective to the agenda of the EEC: the independence of Europe vis-à-vis the United States. With Great Britain, Denmark, and Ireland joining the EEC in 1973, another set of dynamics was added to the configuration process. Further additions of countries to the community through the 1980s and 1990s added other issues to it. By this time, there was a greater emphasis to have supralevel agencies or European institutions with broader powers such as the European Commission, the European Council, and the European Parliament. From this, the EEC was transformed to the European Community with a common market for capital, goods, services, and labor. This was then a move toward more political integration beyond the economic whereby certain parts of national sovereignty gets ceded to the community.

The intense move toward economic integration and political supranationality was conditioned further with the collapse of the former Soviet Union and the unification of the two Germanys. The latter provided certain dynamics whereby with unification of the two Germanys, the German economy by then would have at least 30 percent of the GNP of the European Community. Based on this, there was increasing interest to push for further integration of the economies with that of Germany's in view of the historical memories of Germany's past. What it led to was a single European currency and the formation of an independent European Central Bank. Other institutions of a supranational nature added were the Court of Justice and the Court of Auditors. With the Maastricht Treaty of 1991, the stage was set for integration in all aspects from infrastructure, technology, justice, police, environment, immigration, regional development, and a European constitution. The latter has yet to become a reality. The economic and political developments began to make national policies of the respective member countries less determinative. Of course, there were opt-out options for those states that did not wish such an extensive integration.

Localization and Decentralization

Opposite to the regionalization-globalization framework, on the other end of the political configuration scale is the move down the organization scale to that of governance via local communities. Historically, we find this process, for example, in the emergence of the city-states of early Greece. This level of governance is not one that is unique because it has existed in the past and exists today in terms of political governance at the city or municipal-county level, though the political function of such a political setup is rather limited in terms of juridical authority over various facets of everyday life, for most of the authority has been transferred in the past to the federal level, when nationhood was achieved. However, what is proposed via the organizational concept of bioregion is of different emphasis (Sale 1991; Carr 2004; McKibben 2007).

Where contemporary governance via the nation-state exhibits centralization of political authority, for a bioregion decision making is decentralized to the local level, though there is not an explicit specification for national political authority in the public sphere to date—"there is no hint of 'capturing state power'" (Carr 2004, 101). By focusing on the local area or region circumscribed and defined by the ecological landscape, the tendency therefore is to concentrate on the local. Different political relations are likely to be the result as the political domain of governance in civil society is on each specific environmental setting and their own political needs. The politics of regionalization and globalization, which also means the homogenization of political practices and issues as the particular is downplayed or eliminated for the good of the whole, will not find a similarity under the bioregional political framework. The stress is therefore placed on the autonomous evolution of political practices that are compatible to the socioeconomic activities and the ecological integrity of the landscape.

In terms of domains of governance and coordination, the limits can start with local community groups focusing on common issues such as restoring natural systems, satisfying basic human needs, and reinhabitation. From these community groups there is then the potential of forming a watershed council, the latter being the geoecological boundaries of a bioregion. Such political mobilization and coordination underlines the emphasis on horizontal networking and organization via networks. Decision making thus is by consensus.

If coordination is required for bioregional communities, this is handled through bioregional continental congresses. These meetings are for an exchange of life experiences rather than a colloquium whereby rules are established and legitimated. In this manner, there is not a tendency to develop hierarchical structures but the emphasis is on reciprocity and complementarity. Shared responsibility becomes the political relationship and a network type of organizational structure frames the political configuration and process.

Naturally with local governance comes the issue of the economics of bioregional life. Instead of an environment where competition is the basic drive of the economy, bioregional economics is on cooperation, interdependence, and considerate relations between humans and humans with other living species. Local skills, resources, and knowledge are fostered against the trends and tendencies of globalization that we witness today (e.g., see McKibben 2007). Self-reliance and sustainability are the keys to the economics of production. Therefore, recycling, diversification, and the production of food become key to the basis of the economy along with minimization of the throughput of energy (e.g., see Devall 1993). Community exchange, community transportation networks, and community control of investments and sales are also some of the main features of this economic model.

A four-prong strategy of converting to such an economic model has been suggested (Carr 2004). The first strategy is to identify community issues and resources. This requires the reviewing and mapping of patterns of wildlife, the watershed, migration routes, as well as the patterns of human population. There is also the need to determine the level of needs and services that are produced and supported locally. It might also require the identification of community needs that are not being met and need to be initiated. The outcome would be a bioregional mapping and inventories of the watershed or community.

The second strategy following this would be to oppose any system in the bioregion that is not in alignment with bioregional audits. This can be undertaken through campaigns, political activity, and education.

The third strategy would be to initiate and redirect energies away from the old consumer mentality and transform it to a sustainable system through community-based systems such as credit unions, loan funds, cooperatives, and recycling. Changing land-use policies and community-supported agriculture would also be keys to transforming to a self-sustaining economy.

Finally, the fourth strategy would focus on establishing sustainable systems that are in alignment with bioregional principles such as markets for local industries, food production and exchange, community control of local resources, land stewardship, maximum energy self-sufficiency through alternative fuel sources and balance, community-support housing, and participatory democracy.

Unlike the regionalization project, the bioregional project has not been practiced to the full spectrum of its political characteristics, especially when it has declared that the capture of state power is not its intention, and its area of focus is on civil society in particular. Specific aspects where possible have been undertaken in various parts of North America whereby local communities' attempt to undertake such political projects have occurred at the city or community governance levels (Carr 2004). The adoption of the bioregional political vision by some cities and state agencies in North America has led to some interesting experimentation with the political values of bioregionalism. City planning and community projects are also areas whereby this bioregional governance has occurred. A number of continental bioregional congresses have been held, including one for the Americas.

NOTES

1. For regulative principles and tendencies, the works of Lipietz (1987) and Michel Aglietta (1979, 2007) in discussing the regulatory principles of a specific mode of production and organization that determine the dynamics (and possible crisis) of a social formation—though focusing mainly on the anthropocentric dimensions—have interesting parallels.

2. Paul Volcker, the chairman of the U.S. Federal Reserve Bank under President Jimmy Carter, changed U.S. monetary policy from a long-standing commitment to Keynesian fiscal and monetary policies with full employment as the key objective to a policy in favor of quelling inflation regardless of what it does to employment levels in the United States.

3. In other words, in some places of the world where the technological-digital divide is extremely wide, such as in Africa, parts of Central and South America, and some countries in Asia, this restructured technological-economic space and the opportunities that come with it might not be as available (Castell 2001; James 2003; Wilson 2004; Kagami et al. 2004).

4. The Athenian League or the Delian League was a confederacy of Greek city-states founded to combat Persia. The members, including Athens, undertook military operations against Persia, and thus extended the league's control to the coast of Asia Minor. Financial and military assets were contributed by each member to the league.

5

The Futures

System transformations occur over the long-term as a result of adaptations to a changing environment. No doubt these adaptations by the social system occur either in reaction to shifting conditions and/or as the result of actions conduced by material conditions of life. In the prior chapters, the trends, conditions, and tendencies of the natural and social systems following long-term intense relations between the systems were identified, and the emerging structural changes in the social system were depicted and characterized. What these emerging structural changes suggest is a social system in a process of reconfiguration with multiple transformative paths. It is an evolving process of adaptation to a changing natural environment and climatic cycles and the socioeconomic and political conditions of the social system's landscape. This means that transition to a reconfigured system is a long-term process, and the transformations will take centuries to set and, of course, will be shaped by the conditions of the socioeconomic and political landscapes and the struggles or compromises reached.

What are the features of this reconfigured system? We have distinguished some of them in the previous chapter. Rather than being speculative, we will utilize prior systems transformations (in the form of Dark Ages) as our guide in mapping out further these features and configurations so we can envisage ecological futures that can be based on material conditions of the past.[1]

According to our analysis, the First Dark Age crisis (2200 B.C.–1700 B.C.) was one whereby the system crisis did not produce specific structural transformations but rather specific shifts in geoeconomics and geopolitics in terms of the economic, political, and geographic hegemony moving from one region of the world system to another. The Second Dark Age (1200 B.C.–700 B.C.) crisis was one that was system transformative for structural changes occurred as we delineated in chapters 3 and 4. The Third Dark Age (A.D. 300/400—A.D. 900) crisis shares similar characteristics to that of the First Dark Age crisis. There were no structural transformations, only geopolitical and geoeconomic changes. All of these patterns were outlined in my two earlier monographs: *World Ecological Degradation* (2001) and *The Recurring Dark Ages* (2007).

What is clear is that systems transformations occur very rarely, and when they do, these changes take a long time to complete and set. Multiple paths emerge during the transitions and new structural elements emerge to address the changing natural and social conditions. No doubt the structures of the past will continue as well, for history is one whereby the baggage of the past is carried forward as human social systems do not immediately and easily dispose of institutions, ideas, and practices, especially for hierarchical systems, with their accumulated surpluses and power relations at stake (e.g., see Snodgrass 1980). Multiple trajectories thus are followed and tried depending on the state and level of resources (natural and social) available, and the existing socioeconomic and political conditions. In the long run, as with past historical social systems, the structures that can meet the new contingencies remain while those that cannot dissipate into the dustbins of world history.

What follows is an attempt to anticipate our possible ecological futures based on the historical patterns of the past. In other words, let history be again our teacher of life!

THE AGE OF SILICON

Late Bronze Age Crisis and Late Iron Age Crisis

Earlier we delineated the structural changes that occurred in the late Bronze Age Crisis (Second Dark Age) from 1200 B.C. to 700 B.C. that transformed the system and ushered in the Age of Iron. Using these historical, transformative structural features as a guide, possible ecological futures for the long term can be drawn, assuming certain historical logics of the accumulation process continue uninterrupted. In chapter 4, we identified the main structural transformative changes that were in place at the end of the Second Dark Age crisis. These structural changes developed to be the organizing principles of the newly transformed system that circumscribed the dynamics of production and reproduction of the historical social system then. Using these identified structural features, can we then map the possible structural changes that can be anticipated for the next Dark Age (Fourth Dark Age–Late Iron Age Crisis) assuming that parallel forms of historical transformation do occur? Listed below, and not necessarily in chronological order of appearance and transformation is what we anticipate (though not historically realized) as the organizing principles of the historical social system that could result:

Structural Change I
- Second Dark Age–Late Bronze Age Crisis: Replacement of base material from bronze to iron.
- Fourth Dark Age–Late Iron Age Crisis: Replacement of base material iron-steel to silicon-carbon based.

Structural Change II
- Second Dark Age–Late Bronze Age Crisis: Increased productivity of iron implements led to changing social relations.
- Fourth Dark Age–Late Iron Age Crisis: Adoption of computer-based infrastructure and networks leading to expansion, global production, and wider access.

Structural Change III
- Second Dark Age–Late Bronze Age Crisis: Shift in economic structures from command-centered palace economies to mercantile systems.
- Fourth Dark Age–Late Iron Age Crisis: Shift from market-determined models to ecology- or sustainability-determined models.

Structural Change IV
- Second Dark Age–Late Bronze Age Crisis: "Polis" development leading to new political structures in Greece.
- Fourth Dark Age–Late Iron Age Crisis: Regionalized political structures and localized network-based communities.

Inspecting the above structural transformations for the (next) fourth Dark Age (Late Iron Age Crisis), I note that some features of the structural transformations are occurring, and their continued structural adaptations and evolutions are conditioned by existing natural and climatic conditions and by social, political, and economic struggles and challenges. As the future is still open, what are the various destabilizers, through their articulations, likely to condition the direction of transformation and adaptation?

STORM CONDITIONS

I: Climatic Perturbations

Notwithstanding political violence, wars, and struggles that usually erupt in the course of world history over control of critical natural resources and/or economic and political competitive acts of supremacy, these acts of violence will continue to make their presence felt throughout this millennium. However, the one world-encompassing crisis that systems will face is the impending projected global warming. Coupled with this global danger is the anticipation of increasing scarcity of fossil fuel that will begin to show a decline in world supply by 2050. Together, they will further compound crisis conditions. The decreasing supply of fossil fuel will definitely add to political and economic tensions if it has not already done so in the new millennium.

The Intergovernmental Panel on Climate Change (IPCC), in its Fourth Assessment Report, has stated that there is a 66–90 percent probability that global temperatures will rise between 1.8° and 4.1°C by 2100 (IPCC 2007, 823). By year A.D. 3000, this temperature elevation will be in the warming range from 1.9°C to 5.6°C. What this means is that what we emit in the twenty-first century will be experienced into the next millennium. An indicator of this impending global warming is that the temperatures in the Northern Hemisphere for the past fifty years, according to the IPCC, have been higher than any fifty-year period in the past five hundred years. The alarm sounded is even worse than we can imagine: according to the IPCC, the past fifty-year temperatures in the Northern Hemisphere were the warmest in the last 1,300 years. With these temperature patterns, even if greenhouse gases hold at year 2000 levels, temperature increases will still be about 0.1°C per decade in this millennium. More increases are predicted if the levels of greenhouse gases go beyond year 2000 levels.

The source for these elevations is, of course, anthropogenic in origin. The three greenhouse gases (carbon dioxide, methane, and nitrous oxide) have reached levels far beyond the 1750 preindustrial level. The sources for these gases are fossil-fuel use for carbon dioxide, while those for methane and nitrous oxide come primarily from agriculture. Both sources are primary drivers of modern lifestyles. The concentrations reached for these gases have surpassed prior durations that are millennial in length. The global atmospheric carbon dioxide concentration has reached a value of 379 ppm^3 in 2005 from 280 ppm^3 in 1750. The level reached has surpassed the natural range for the last 650,000 years (180–300 ppm^3). Atmospheric concentration of methane has increased to 1,774 ppb in 2005 from a level of 715 ppb in 1750. The level of methane exceeds the natural range of concentration for the last 650,000 years, growing from a level of 320 ppb to 790 ppb. Nitrous oxide atmospheric concentration has also risen to 319 ppb in 2005 from a preindustrial value of 270 ppb.

It is well understood that all these increases have other consequences in the realm of Nature. The temperatures of the world's oceans have increased to at least depths of 3,000 m, which means that the oceans have been absorbing 80 percent of the heat added to the climate system (IPCC 2007). Because of these temperature increases, mountain glaciers and snow cover have declined in both hemispheres, which means, in some cases, rises in sea level and loss of water resources. In addition, anthropogenic forcing has also contributed to changes in wind patterns, thus affecting storm tracks and temperature patterns. All this will result in more violent hurricanes and typhoons.

What are the implications of these trends and tendencies as a consequence of global warming and the effects of after "peak oil" on the reproduction of social and natural systems? In terms of impacts and vulnerabilities that the social system will have to address, they are quite far ranging. Projected temperature elevations will impact the regions of the globe differently.

In the Sahel region of Africa, the temperature increase will lead to drier and warmer conditions, an impact on growing seasons that will adversely affect agricultural harvests. Rain-fed agriculture in Africa will see a decline in yield of up to 50 percent. However, for North America, aggregate yields from rain-fed agriculture will increase by 5–50 percent (IPCC 2007)! Generally, regions in mid- to high-latitude areas will find a slight increase in crop productivity, while those in the lower latitudes and in tropical regions will see the opposite condition arising.

Elevated temperatures will result in the crisis of water availability. With water stored in snow cover and glaciers projected to decline, water from these sources is projected to be reduced to the extent that it will impact at least one-sixth of the world's population (IPCC 2007). In other parts of the world less dependent on snow melt, like parts of Africa, an estimate of between 75 million to 250 million persons will experience water stress by 2020. In Central, South, East, and Southeast Asia whereby freshwater availability is projected to decrease in the river basins as a consequence of climate change, a billion persons will be impacted by the 2050s (IPCC 2007). For Latin America, elevated temperatures and decreases in soil water will lead to replacing tropical rainforests with savannas in eastern Amazonia, besides deforestation that is anthropogenically engendered. Salinization and desertification of agricultural land will also follow as a consequence of climate change. North America will experience conditions of reduced summer flows due to decreased icepack and more winter flooding.

Climate change will also cause sea-level rises that have tremendous impact on coastal areas. In places like Asia and Africa, where the capacity to deal with this rise in sea level is not as robust, the impact will be devastating by 2080. For North America, with high-value infrastructure and properties, climate variability will mean an increase in the intensity of tropical storms, thus causing severe economic pressure on economies.

Global warming has various impacts on health. We will find the health status of the global population affected by an increase in malnutrition because of harvest failures; increases in deaths and diseases due to heat waves, floods, storms, fires, and droughts; diarrheal diseases on the rise as a consequence of flooding; and also an altered spatial distribution of some infectious disease vectors. Especially for rise of diarrheal-disease incidences, East, South, Southeast, and South Asia will be impacted by global warming.

Systems Reaction to Climate Change

Clearly the effects of global warming as described above will engender adaptive measures to meet the contingencies and challenges as the global economy will be impacted severely. Initial projections of economic costs that will be incurred by climate changes are quite alarming. For the core areas where the economic structures are the most embellished, the levels of economic losses will be extremely stressful on economic reproduction capacities. Just in terms of extreme events such as storms, floods, droughts, and heat waves, such disasters have been estimated to economically cost about 0.5–1 percent of the world's GDP by 2050 (Stern Review 2007, 131). Such economic damages have been occurring since the 1990s, and in 2005, the costs were about $200 billion—more than 0.5 percent of world GDP of that year. On the average, since the 1990s, it has cost about 0.2 percent of world GDP. Payouts by insurance companies for hurricane-related damages in 2004 and 2005 amounted to $145 billion and $200 billion respectively (*Economist*, September 9, 2006, p. 12).

For example, if we translate down to the nation-state level, in 2005 Hurricane Katrina's costs totaled $125 billion, amounting to about 1.2 percent of the GDP of the United States. Table 5.1 outlines major natural disasters costs for the United States from 1992 to 2004. The European heat wave of 2003, when the temperature rose above 2.3°C, long-term average costs were $15 billion in damages to farming, livestock and forestry. Human costs totaled about 35,000 deaths.

Table 5.1. Costs of Major Natural Catastrophes for the United States, 1989–2004 (Billions in 2004 Dollars)

Date	*Event*	*Insurance Payouts*
September 1989	Hurricane Hugo	6.39
August 1992	Hurricane Andrew	20.87
January 1994	Earthquake (Northridge)	15.93
September 1998	Hurricane Georges	3.36
June 2001	Tropical Storm Allison	3.1
August 2004	Hurricane Charley	7.48
September 2004	Hurricane Ivan	7.11
September 2004	Hurricane Frances	4.6
September 2004	Hurricane Jeanne	3.66

Source: New York Times, September 3, 2005, B1.

Initial projection of a 3°C increase in global temperature will mean a 5–10 percent rise in the intensity of major storms for the United States that will likely result in total losses of 0.13 percent GDP each year, amounting, on average, insured losses of $100 billion–$150 billion a year. Estimates of storm damage for Europe per year will rise to $120 billion–$150 billion by the end of the century.

Sea-level rises will not only affect rural coastal communities, as mentioned previously, but also some global cities (London, New York, Tokyo) will also be impacted. For example, possible damage to physical structures for the city of London has been estimated to be about $220 billion of assets that are lying on the flood plain (Stern Review 2007). In view of the scale of such monetary costs, insurance companies will need to reserve capital to cover the extreme losses, which means the costs of capital borrowing for other investments will increase automatically, thus further impacting economic growth.

In the periphery of the world system, economic loss will be less as the level of investments in infrastructure is not as high as the core. Economic losses will be what the Stern Review (2007) terms *development losses*. In South Asia, for example, changes in monsoon cycles would mean that agricultural production in northern India could be reduced by as much as 70 percent by 2100. Against this drop there will be the need of an additional 5 million tons increase in food production each year to keep pace with population increases that will reach 1.5 billion by 2030. Sub-Saharan Africa will face hunger whereby 250 million–550 million people will be at risk. Of Sub-Saharan Africa's plant species, 25–40 percent will increasingly find an inhospitable habitat by 2085 because of temperature increases. Latin America also seems to share the same type of tragic conditions with global warming. The Amazonia will be reduced, and by 2050, 50 million persons will be at risk in terms of water access. In Asia, China, the most populous country, will see a decrease in agricultural farmlands amounting to about 25 percent of arable land, thus reducing by 14 percent the value of total agricultural output.

Such challenges of a world systemic nature will definitely result in socioeconomic tensions causing migrations, wars, and famine, notwithstanding the losses to the world economy and respective national or regional economies. These tendencies add on to historically existing sociopolitical conflicts that have been in play, except that in this case, the origins are derived from climatic changes.

II: Peak Oil

Compounding already precarious ecological conditions, the scarcity of oil will have severe consequences for the global economy. Calculation of discovered reserves of crude oil in the world is expected to peak around 2010 but no later than 2020 (Rifkin 2002; Meadows et al. 2004; Brown 2006). By definition, peak oil is defined as the point in time when about half of the estimated recoverable reserves of oil in the world have been produced. Projected resource life expectancy with year 2000 as the base line is estimated to be between fifty to eighty years (Meadows et al. 2004). Dropping down to the level of national reserves, the projected long-term outcome at the start of the new millennium is dire. The lower forty-eight states of the United States are estimated to have about 195 billion barrels of recoverable oil of which 169 billion barrels have been extracted, leaving about 20 billion barrels and perhaps about 6 billion barrels that could be found, most likely in Alaska. Russia has about 200 billion barrels of recoverable reserves but has pumped up 121 billion barrels, leaving only about 66 billion barrels in the ground and perhaps 13 billion barrels to be discovered. Saudi Arabia, which the West highly depends on for its oil, has recoverable reserves of 300 billion barrels of which 91 billion barrels have been extracted, leaving about 194 billion barrels. The possible yet-to-be-discovered number of barrels is 14 billion.

Given such a global scenario—especially if alternate fuel sources (nuclear, wind, solar, hydrogen) cannot meet global demand as a result of technological obstacles or the lack of political will to make the necessary transitions as a result of the strength of class hegemony, domination, and control—energy shortages can lead to undermining the current globalization process, which entails global production chains that use tremendous energies to produce and circulate globally manufactured commodities. Any interruption to this global process will mean a crisis of realization for the world economy and, hence, the socioeconomic and political orders. It is very likely that wars and violent measures will be taken to secure energy supplies by those regions and countries at the top of the hierarchical power structure. This has occurred in the past as we have shown in chapter 3, even as far back as five thousand years ago when wars were undertaken to have a secure timber supply, and during World War II. More recently, some have even suggested that the Persian Gulf wars have been fought to ensure that the advanced, developed economies do not have their oil supplies interrupted. Why should it be different now and in the future? Even the People's Republic of China has already shown signs of intentions to increase its access to energy supplies by signing energy contracts in Africa, Canada, and other parts of the world. Chinese foreign aid in friendship to Africa is also another avenue of ensuring access to needed natural resources and energy supplies.

The decrease in fossil-fuel supplies for some can be met by technological adaptation and innovation in terms of alternative energy supplies, more efficient use of remaining fossil-fuel sources, and perhaps conscious changes in lifestyles. The transition to any of these alternative strategies remains a huge challenge in light of the socio-organizational patterns of a hierarchical world and its associated hegemonies.

III: People, Places, and Practices

When Paul Ehrlich published *The Population Bomb* almost four decades ago, he warned of the consequences that runaway population growth has for our planet. Since its publication, the world's population has reached 6.6 billion according to 2007 estimates of the U.S. census. The exponential growth is dire especially when the world's population was only 5.25 billion in 1990 (Meadows et al. 2004). Depending on different circumstances, the low projections for future world population growth by the United Nations—assuming a rapid drop in fertility occurring nearly everywhere—is expected to be around 7.5 billion in 2035 and falling further to 7.4 billion by 2050. If a medium-growth tendency is assumed, the world population will reach 8 billion in 2025, and about 8.9 billion in 2050 (United Nations 2003). Clearly, the low projection is too optimistic, especially if at current levels we are already over 6.5 billion.

Of course, the trajectory of fertility rates for Third World countries, which will likely be impacted by diseases, such as AIDS, diarrheal inflictions, SARS, and so forth, have not been fully realized. Already, for example, we find that seven countries in southern Africa, where the infection rates of AIDS have reached 20 percent or more, have been projected to expect little or zero population growth by 2050. The other major possible reducers of global population growth are SARS and the avian flu diseases that most likely will start in Asia, where the viruses have consistently appeared, and will reduce world population growth should pandemics occur. Given that Asia is the zone whereby some of the world's most populous countries are located, it will mean major drops in existing population levels and decreasing fertility patterns for the long term. Furthermore, with the globalization of air travel, the spread of such diseases could be worldwide thus affecting the population growth rates of the world. The possibility of such events occurring should not be discounted nor dismissed (Chanda 2007). The United States, for example, has already established quarantine protocols should such pandemics occur (Homeland Security Council 2006).

Depopulation therefore is just as much on the horizon as population increases. Mixed into the mélange of demographically reductive conditions that are likely to be caused by global warming effects and wars, diseases therefore can be a potent addition in terms of depopulation in the future. As we have witnessed in the prior chapters, plagues, cholera, and influenza were socially and economically disruptive in world history and especially during the previous Dark Ages. Given such circumstances, regardless of whether low or medium population-level projections are reached, and discounting for population losses due to diseases, global warming, and other social causes, the level of the world's population as Meadows et al. (2004) have projected will pose tremendous barriers toward sustainable living in light of diminished resources. Therefore, population will continue to be a destabilizing force impacting the system and conducing conditions of crisis demands in already depressed social and ecological environments.

When Gordon Childe (1942) proclaimed during the last century that the Urban Revolution was just as world historic as the Neolithic Revolution, it was assumed by most—especially in light of the urbanization process over world history–that this human endeavor was not going to end and that it would continue to grow without any limits. This has continued to be the impression for most urban demographers and it is echoed in the recent projection by the United Nations Population Fund (UNFPA) (2007). The declaration is that for the first time in world history, more than half of the world's population (3.3 billion) lives in urban areas. The prediction is that by 2030, the urban population will increase to 5 billion.

Whether the urbanization process will continue limitlessly is a major question. We have shown that this has not been the case historically, especially during Dark Ages when deurbanization has occurred. The United Nations Population Fund, in a 2007 singular report, has offered conflicting trends. It pointed out that even though overall urban growth rates has consistently declined in the latter half of the last millennium in one section of the report, it insisted throughout the report that urban growth continues to be a threat to the human condition in the Third World. In fact, UNFPA has termed this growth *the second urban revolution* with the urban revolution starting in the late eighteenth century in the West as the being the first in world history! How do we make sense of this projection and assessment that to us are historically myopic and modernistic? In the short term, the growth in urbanization is expected. If this is the case, an increase in resource consumption is expected as urban areas consume about 25 percent more than rural areas (Brown 2006). Such a tendency will mean an impact on the already stretched resources available globally thus ratcheting up all the trends (access to clean water, employment, housing, etc.; see UNFPA 2007) that will lead to greater social disparity and hence social unrest.

In the long run, however, if past historical tendencies are repeated, deurbanization will occur, if it has not already as we indicated in chapter 3 (e.g., see Ostwalt and Rieniets 2006). Even the UNFPA (2007) reported that urban growth has declined in the last millennium.

The above trends and tendencies will likely mean the continuation of the global migration of peoples from socially and environmentally threatened areas to "lands of gold." Worlds in motion are the order of the day as we have outlined in chapter 3. Global movement will continue to heighten as socioeconomic disparities mount as a consequence of various social, economic, political, and environmental policies that regulate the world's economies and political arenas. Global migration will raise primordial issues of racial and place identities, and political and juridical rights of access to the resources of the land on which the communities have their abodes. As we have stated in considerable detail in terms of the tensions and pressures on sociopolitical stability in chapter 3, migration again appears as a transformative process just as it did in prior histories.

Increasingly among scholars of globalization, place identity known in the contemporary era as citizenship has been undermined with increasing globalization and regionalization whereby expert expatriates and legal migrations continue to take place (Ong 2006; Sassen 2006). With a world on the move, the issue of the disaggregation of citizenship into different types of rights in order to be able to incorporate noncitizens has been raised (Benhabib 2002). With limited benefits packages and civil rights for migrant workers provided in some states in Europe, such policies will lead to what can be termed *partial citizenship* (Soysal 1994). As we indicated in chapter 4, with advances in information technologies, virtual migration as part of the global division of labor is also being instituted (Aneesh 2006). This will further diminish the privileged status that place identities, such as citizenship, in the past offered in terms of access to high-paying work, especially for those located in the core of the world system. The regionalization process occurring in Europe will further strain this issue of the privileged status of citizenship, as labor barriers and technical-skills criteria are removed and homogenized to mesh within the single economy of the European Union. In other words, those persons who used to live in economies of the EU that have lower wages (such as the former Eastern European economies that have now joined the EU) will have the opportunity to compete for higher wage jobs in other parts of the EU.

REALIGNING HOME

For a system organized on a hierarchical distribution of capital surplus within and between territorial boundaries, those at the top of the hierarchical distribution, as past practices have shown, will attempt to maintain the status quo in reaction to whatever adaptive measures are taken in response to the global environmental crisis and global warming. Therefore, the least system transformative route that will not undermine power relations and profit realization will be the most likely step followed. Thus, the adoption of better sustainable technology and better ecological management of available resources have been consistently promoted, and we can see this in the current era where steps have been taken to deal with the impending crisis without changing the organizing principles of the social system.

Market optimism, regionalization, and globalization policies and practices will be pursued until ecological and natural limits are reached. The "business as usual" approach will be fostered similar to what we witnessed in the palace-centered kingship economies that persisted at the end of the Late Bronze Age crisis (the Second Dark Age). No doubt, as the catastrophes continue to mount as effects of global warming compound and recur, more stringent measures will be implemented to maintain economic, political, and social control. Barbarism that is often connected with system collapses will likely emerge in the form of rules of governance and political and military actions, if it has not appeared already. One can anticipate the formation of "fortress" type configurations to keep those within safe and those that are outside out. Early signs of such actions can already be seen in the enactment of immigration policies and laws, whereby legal and illegal immigrants are screened and blocked.

Furthermore, with energy shortages it is very likely that certain places in the world will become more isolated. By no means should this be seen as negative. Like the monasteries following the collapse of the Roman Empire, isolation can provide the opportunity for innovations as the predominant or common way of managing socioeconomic and political affairs in the globalized world no longer can be practiced or is an option. In this case, centralization is no longer a practical affair. Localization becomes the only possibility. Such an environment can only arise in the context whereby environmental limits have been severely exceeded and socioeconomic and political chaos abounds. It is only reached during the depth of Dark Age conditions such as the Second Dark Age crisis has shown. Current conditions do not indicate that we are at that stage yet. What then are the potential trajectories in the two primary spheres (political and economic) that are critical in the reproduction capacities of the social system?

THE POLITICS

Beyond Westphalia

The Second Dark Age crisis (1200 B.C.–700 B.C.) engendered structural changes of which the widespread utilization of iron as a base metal and the emergence of a political framework known as the *polis* became systems transformative. With these two structural shifts, transformations of social, economic, and political relations and ideas followed concomitantly. But such structural changes brought neither an end immediately the utilization of bronze in material life nor the disappearance of the palace-kingship political pattern of the past—the political organizational configuration that had been in place with the formation of urbanized communities in world history five thousand years ago. In fact, following the end of the Second Dark Age crisis, these palace-centered political and economic systems continued, though increasingly their control of the economy was undermined by the use of iron that they did not monopolize—in terms of trade and manufacture—in comparison to what they had controlled when bronze was the main metal of material life (e.g., see Snodgrass 1980, 28).

Politically, therefore, the kingship pattern continued in the eastern Mediterranean and the Near East, with the exception of the cities in Greece. Polis-type governance became a more predominant political practice later. It is only with the establishment of Greek colonies all over the Mediterranean and the Near East from 700 B.C. onward, the ascendance of Rome, and Roman adoption of Greek political concepts that a political system based on republican rule—though interrupted at times in the Roman case by the inauguration of the emperor—became more widely practiced.

In comparison, for our current *conjoncture*, we are already starting to witness, as described in chapter 4, the emergence of the political process of what we term *regionalization*: the union of independent nation-states, initially starting with trade treaties and finally extending to political union (and later perhaps with a common constitution), as we are beginning to witness in Europe with the formation of the European Union (EU). Coupled with the economic process of the current era's globalization strategies, the regionalization initiative as a political framework will be facilitated by a set of suprastate structures such as the International Bank for Reconstruction and Development Group (World Bank), the International Monetary Fund (IMF), the World Trade Organization (WTO), and a series of regional banks, for example the Asian Development Bank (ADB), the African Development Bank (AfDB), and the InterAmerican Development Bank (IDB). Besides Europe, the other regions of the world have not evolved as far in terms of political regionalization. In Asia and the Americas, the regionalization process has been mostly focused on economic integration and trade exchanges. Hence we find the Association of Southeast Asian Nations Plus Three (ASEAN Plus Three) compact and the North American Free Trade Agreement (NAFTA), for example.

A regime of control through monetary arrangements and free-trade declarations via the auspices of the most powerful advanced industrialized countries, some of which are part of the EU, ensures that global economic order is maintained. The articulation of regionalized political configuration with the current setup of nation-states as the organizing principle by which global political governance and political order are exercised and legitimated has yet to gel. Besides this, what does this regionalized reconfiguration mean for the "old" concept of hegemony via the nation-state with the United States as the declining hegemon in a unipolar or multipolar world? Does it still reflect the political economic reality or the notion of empire in the form of a "power network" with no center as depicted by Hart and Negri (2000) is a better depiction of global relations? It is perhaps too early to judge. But what is clear at this point is that a structural form (regionalization) has emerged in the European Union at least, though its full development has yet to be realized.

On the other end of the spectrum is the increasing interest in some parts of the world with the political concept of *localization* in terms of political governance. Primarily spurred on by the effects of the technologization and centralization of socioeconomic and political life that have been going on since the eighteenth century and reaching a crest with the current ongoing globalization process that destroys local identities and homogenizes sociocultural lifestyles across spatial boundaries, localization has gained some momentum as a reaction to the threatening processes and trends of globalization and the consequences of the destruction of Nature and the modification of natural processes (e.g., see Hardt 2002; Carr 2004; Aberley 1998; McKibben 2007). Such calls for local governance and control unfortunately rub against the grain of the more dominant push for regionalization and globalization, for the latter underlines the regime of control that is in place at this point in world history. As mentioned in chapter 4, some in-roads have been made mostly at the county, city, or community levels whereby control of the governing process is easier to win. In addition, there are also other signs of decentralizing in terms of initiatives at the state level by some U.S. states to deal with environmental and other issues that are not being addressed by the federal government. For example, initiatives to control global warming and stem cell research at the state level have been passed by several state legislatures.

The shift to such a political reality where locality instead of regional and global systems of governance becomes the norm would mean, from a modernist position and social-evolutionary complexity standpoint, a devolutionary step. If viewed from a modernistic worldview, it is a regression—a diversion from the "progressive" move to universality. It is a tribal-communal configuration—a political orientation of the past with less hierarchization. But such developments are not unusual in world history. They have occurred before with studies from northwestern Europe showing the various times of such political reconfigurations during periods of the Dark Ages (Kristiansen 1998a, 1998b, 2007). Though it might seem unlikely—especially with the current ongoing globalization putsch—that such a political transformation can become a predominant political framework, circumstances might change that can force it to be so. It has occurred in history—during the Dark Ages!

For the foreseeable future, these two main options (regionalization and localization) of political processes—there might be other paths as well—will continue to develop paralleling what we witnessed at the end the Bronze Age whereby palace-centered kingships and polis-type city-states continued to politically adapt to the changing conditions of the time. Of course, a combination of political paths (regionalization and localization) could be taken in different parts of the world leading to different outcomes depending on the circumstances. This was the trajectory that was taken during the Dark Ages at the end of the Bronze Age. Such continuance and evolution of current political structures can be interrupted or arrested by extreme crisis conditions that will likely emerge from the effects of climate changes, diseases, natural resource scarcities, and, of course, socioeconomic and political dynamics. Therefore the political structures (regionalization, localization, or a combination) that are in the process of "becoming" will be conditioned by such circumstances.

What could likely tip the shift toward a political-localization type of governance whereby central control no longer can be exercised? As it occurred in previous Dark Ages, the tipping points would most likely be the increasing scarcity of natural resources and climate changes that would undermine the reproduction of existing lifestyles and consumption. Such restrictive parameters will force the affected areas to turn to practices that can reproduce life under local conditions for the old ways of consumption no longer work. Politically, therefore, central authorities no longer could provide the resources to meet contingent needs, and as a result, the legitimacy of the central authorities in terms of governance is severely undermined. This undermining process is compounded further with social movements resisting the prevailing conditions, and the latter often comes into play especially during devolutionary periods.

Social Movements as the Angel of History?

Whether active human agency will play a significant part will depend on the contingent structural conditions of system. Can we confidently state as what some "progressives" have asserted on the dynamic role that social movements (e.g., the World Social Forum) will play in shaping the evolution of the social system (e.g., see Wallerstein 2005; Hardt 2002; Klein 2001)?

Transforming the system via social movements or classes has been on the theoretical and political agenda since Marx called for the workers of the world to unite. This historical angel of history has been considered as the force that will make things right. Over history, this angel has been renamed, recomposed, and realigned according to the political and socioeconomic conditions of the times (e.g., see Gorz 1982). In the late twentieth century, the angel has been recategorized and termed as *social movements*, devoid of the restrictiveness so that others beyond the working class can be included as a transforming angel of history, thereby ensuring inclusiveness. Therefore, any organized resistance from the Zapatistas of Mexico to the Assembly of Poor in Thailand can be considered as part of the family of social movements of resistance—the multitude (see Hardt and Negri 2004). As such, most of these struggles have local roots and occur all over the world, and, according to some, they all form part of an expansive struggle to the inequalities of what globalization has generated (Klein 2001).

The crystallization of these varied movements in the new millennium resulted in the formation of the World Social Forum in Porto Alegre in 2001. "Another World is Possible" is its rallying cry. There has been much discussion on what the overall mission of the World Social Forum should be. These have ranged from those calling for reform of the system to those who want to replace it. Likewise, there are some who want to negotiate with the globalization forces and those who seek mobilization and confrontation (e.g., see Petras 2002). Hardt (2002) has even compared the World Social Forum to another major historical event—the Bandung Conference—in the mid-twentieth century when nationalist political leaders met in Bandung to discuss strategies to counter the then-world order. At that time, the nation-state with its territorial "sovereignty" was viewed as the counterforce to the then hegemonic powers.

However, things have changed somewhat. With the globalization tendencies of today where "deterritorialization" is systematically eroding a nation-state's ability to act, we have the multitude as its replacement to challenge current hegemonic powers. Porto Alegre as Hardt depicted it "was populated by a swarming multitude and a network of movements. This multitude of protagonists is the great novelty of the World Social Forum, and central to the hope it offers for the future" (2002, 112). However, we should note that the varied membership of the World Social Forum, ranging from environmentalists to social activists to intellectuals, can further compound the directions of the Forum. If we consider the discussions at Porto Alegre as providing us with an opportunity to assess the World Social Forum's effectiveness, the jury is still out. Take, for example, the issue of political action along national liberation lines or a democratic globalization approach toward challenging current hegemonic powers. The first approach is to reinforce the sovereignty of nation-states as a defense to foreign and global capital. The latter is to strive toward a nonnational alternative whereby any national solution is deemed ineffective, and the only avenue is to take a truly global response. The political differences across these two approaches reflect the myriad of social movement groups that are part of the Forum, thus showing the difficulty of reaching a consensus. Just how effective this global movement can be remains to be seen (see Robinson 2004; Klein 2001).

THE ECONOMICS

Going Green and Virtual

With growth as the basic principle of modern economies, this tenet has spurred incessant economic expansion and unsustainable consumption consequently leading to the depletion of nonrenewable resources and the degradation of the landscape. From this precept, the open "cowboy" economy to borrow a term from economist Kenneth Boulding (1966) has been the dominant theme of social life. With the increasing scarcity in natural resources, the questions posed these days by some economists and experts are whether perpetual growth as expounded by market-based modern-day economics is possible for the long term, and whether there is the need to consider that the world's economy is a part of a larger finite and nongrowing ecosystem (e.g., see Daly 1973, 1996).

Besides the above issues, the other concern expressed is that the "science" of economics should include the inputs from Nature and the outputs of production (besides commodities), such as pollutants, in the overall calculation of the economics of production (see McKibben 2007; Hawken 1999). With a strong faith in technology and rationality, a new age of *green capitalism* is envisaged. The argument is that capitalism is not capitalist enough. The complaint has been that a number of inputs and outputs of the production process are not factored in at all. For example, what is normally treated as ecosystem services (such as pollination and decomposition) and externalities (such as clean air and water) are considered as gratis and exclusive of costs. They are not included in the calculations of overall production costs, for these elements are borne by public expenditures instead of private businesses that normally undertake the production of commodities. Cost estimates of such services provided by Nature have been estimated to be as high as $33 trillion annually. From an economic standpoint, which uses price as a basic element in determining exchange value and market characteristics, the argument by some has been that in order for proper valuation of the costs of producing commodities for incessant consumption, these variables have to be included in the calculations. The aim, therefore, is to get the "price right." It does not mean that on an overall basis there has been a change in assumption that the overall macroeconomy is part of a larger ecosystem with finite resources. Rather, it means that with the inclusion of these variables the actual costs for the production of commodities and hence consumption costs would be much higher. What this means is that for continuous growth besides tangible costs there is the need to take into account the value of Nature and natural processes—Natural Capital as Hawken (1999) has defined it—in the overall production process. The value added by Nature needs to be valued equally with the value added by capital and labor. It is hoped that from such considerations the finite nature of available natural resources and the importance of the ecosystem in the reproduction of economic life will be highlighted and that capitalism should be more efficient in its utilization of Nature's resources.

Given the current warnings of depletion of natural resources by core industrial countries, and with China and India on an accelerated pace to catch up with the advanced industrial economies, will the shift be made in light of the above economic determinations that market-determined economic principles become increasingly unrealistic and that a shift in economic mind-set—or as Daly (1996) puts it "preanalytic vision"—be made? The displacement of contemporary economic models by ecologically driven ones has been the hope professed by ecologically minded economists and scholars (e.g., see Daly 1996; Constanza et al. 2007; Brown 2006; Hawken 1999). The extent of such a change is uncertain.

Will the shift only cover economic valuation of natural capital or will it include as well a move away from a growth-based model to one of equilibrium and steady-state? The latter will mean a qualitative shift in our mode of social relations and the basis of our production and accumulation processes. An economy that continues to produce commodities incessantly for exchange will shift to an economy that develops instead of growing incessantly (see Daly 1996). Quantity gives way to quality. It will entail a complete rethinking of the manner in which we conduct our socioeconomic and political lives.

There is a plethora of proposals offered for a shift to a world of sustainability (e.g., see Hawken 1999; Brown 2006; McKibben 2007; Ehrlich 2004; Olson and Rejeski 2005). These proposals have suggested state-policy changes to encourage and engender human consumption patterns, technological innovations and solutions to meet diminishing resources, ways to increase the efficiency of production so that needs can be satisfied, voluntary lifestyle changes so that the growth model is replaced with a sustainable form of living, and a complete rethinking of the manner in which we conduct economic life. To what extent such proposals will take root in the short to medium terms is a good question. During the transition from the late Bronze Age to the Iron Age, two economic models organized the different economies then as we identified in Structural Change III. The possibility for our era would be the continuation of the growth model and the increasing penetration of an ecologically determined model as the Dark Age crisis deepens. This scenario assumes a social system that makes adaptive changes as it continues to evolve in a changing environment. It means a social system that learns and adapts as it develops and has little difficulty in learning. It means a reflective social system.

However, if this is not likely to be the ecological future we face, an uncertain scenario emerges and collapse is likely to follow. In *World Ecological Degradation*, I have shown that the collapse of civilizations is not really a problem of learning (for human societies do learn) that is the issue, but it is *not-learning* that is the problem. In other words, we choose not to learn from our past mistakes. We choose just to show "the past as it once was" as history has nothing to teach us (Shäfer 2007, 3). This problem in the past has been the continuous exploitation of the natural environment without any regard for sustainability. If collapse has been the historical past for a number of overconsuming civilizations, will this be our historical future?

To avoid collapse would mean addressing the anticipated crisis in natural resources availability and the effects of global climate change. Much has been written and more will be forthcoming offering recommendations, ideas, and proposals, as we have elaborated in the previous chapters. There is little need to dwell on these specific details, and I leave that to the various social, economic, and technological specialists. Rather I wish to pursue the strategic issue of social adaptation and evolution that I believe is necessary for our common future. With the past Dark Ages that have occurred, the social system has been able to adapt and evolve primarily through extensive expansion whereby human socioeconomic and political colonization of other regions of the planet enabled the accumulation process to continue unabated. This trajectory no longer is available for our era as the globe is quite extensively colonized with the exception of the Arctic and Antarctic regions.[2] What then could form the basis to meet consumptive needs? In this case, we need to consider Structural Change II as we delineated at the start of this chapter. The structural shift involves information and computerization to provide the conditions for the further reproduction of the social system: a world of virtuality where value is created and consumed as commodities, and most important, dematerialization becomes increasingly a basis of organizing commodities produced and consumed in an organized networked social world (Mitchell 1999).

In a world of natural resource and material scarcities, socioeconomic exchange will continue and material resources will be needed, for after all we are material beings—we must eat to make history. The challenge will be to meet all the other human needs while utilizing fewer materials, physical spaces, and so forth. The virtual world that has been created by computerization (see chapter 4) as epitomized by the Internet where increasingly socioeconomic and political realities produced and consumed have resulted in the generation of different worlds whereby virtual communities exist in cyberspace. The generation of such worlds requires fewer material resources than would have been necessary if such activities occur in a common public physical space. What these worlds do for human sociality remains to be seen. Castells (2000) has called such socioeconomic and political interactions over cyberspace the culture of real virtuality. The availability of different communication modes interconnected provides the opportunity of the "construction of real virtuality" (Castells 2000, 403). Symbols and signs produced and consumed in cyberspace increasingly form the cultural experience and generates both use-value and exchange value. As Baudrillard (1975, 1983) has so elegantly dissected, the logic of signification comprises of a functional logic of use-value, and an economic logic of exchange value besides the two other logics of symbolic exchange and sign value. Given such a construction, in the world of real virtuality people's material and symbolic existence is captured and reproduced in cyberspace, and what is communicated in virtual images or appearances is not just the communication of the experience, it also becomes the experience (Castells 2000). It is substituting signs of reality for reality itself, and going from this, the generation of models of a reality without the origins of reality (Baudrillard 1983). Such simulations can (if not already are) become the basis whereby socioeconomic, political, and leisure needs can be met where the provision of a tangible material product becomes unnecessary: the satiation of needs via images over cyberspace. No doubt *not* all material needs, such as food consumption, can be met through this, but a significant number of needs could be satiated.

The advances in computerization and communication as we discussed in chapter 4 provide the potential whereby socioeconomic exchanges can be realized and accepted via signs and symbols codified in an electronic medium that neither needs to be concretized in a material tangible product nor requires a physical setting for the socioeconomic exchange to occur. For example, instead of producing and printing a book, warehousing it, and later selling it in a bookstore, the book could be "produced and published" in electronic form and stored in a server to be accessed or sold from anywhere in the world at any time. It would be the remaking of the making (Mitchell 1999). In sum, there would be a net dematerialization. For example, an e-mail message that is read does not require paper unlike the old snail mail. Perhaps we will even have the taking over of big physical sizes with miniaturization that is increasingly the case in manufacturing. The most radical move would be to shift more toward an economy that entails little material products or physical structures—a "weightless" economy.

What will be the impact of such developments for human sociality? It is hard to gauge—definitely there will be "Luddites" who will complain and protest (Sale 1995). However, we are at the dawn of a major system transformation: the Silicon Age will provide a vast array of possibilities whereby homes, workplaces, transportation systems, and electronic communication will be reorganized to produce new patterns, processes, and relationships in response to the contingencies we face in the natural and social environments. However, if such strategies do not take root or are undermined by the continuation of "old" ecologically unsustainable economic-social policies that have brought us to the conditions we are experiencing today—recall the command-centered palace economies that continued following the end of the Bronze Age coexisting with the mercantile systems—collapse will likely follow. At this point in world history with the state of globalization, collapse will be worldwide.

HOPE IN A HOPELESS WORLD

Collapse is *not a condition that we should dread*. As my study has shown in *The Recurring Dark Ages*, it can be a rewarding time for the human community and Nature. For the human community, it is a time of experimentation, innovation, and increasing sociality. For Nature, it is a period of recovery from the unceasing assault by the social system. What will this collapse look like?

Past Dark Ages have provided us with some ideas of the social, political, and economic structural arrangements following collapse (see Chew 2007; Kristiansen 2007). The struggle over limited natural and energy sources will undoubtedly lead to conflicts within and between regions of the world system. Local, regional, and global conflicts will likely result. These conflicts tend to be exacerbated during Dark Ages. They are typical conditions.

With very little opportunity for surplus production as a result of the scarcity of natural resources, the flattening of the social hierarchy of the social order on a world scale will likely occur. Regionally, this has occurred in the past in Greece and Scandinavia during the Bronze Dark Ages (Chew 2007; Kristiansen 2007). Furthermore, with the increasing natural catastrophes (typhoons, hurricanes, El Niño) that are anticipated with climate change along with the spread of diseases, population losses will mean less consumption of limited resources. I expect to see global population decreases due to these types of conditions, and also due to wars and other conflicts. This has happened in previous Dark Ages also. From humanity's point of view, this would be a great loss. But such overall population loss will mean less demand on Nature's resources.

Scarcity of energy will lead to major restructuring of the economy and the polity. The lack of available energy will mean that transportation between locales will be increasingly curtailed, leading to isolation. Isolation will lead to the resurgence of local economies and localization of political governance and control as centralization and urbanization can no longer be sustained. The latter (centralized systems), as we have indicated in previous pages, will increasingly be on the decline as the lack of energy sources will restrict the transportation of food and services that have been the organizing feature of the reproduction of centralized systems and urban life. The deurbanization process (shrinkage) that I have presented in chapter 3 will be further exacerbated. Mass migration will result, and in this case, it is not just for employment but for survival. There will be casualties from such a process. It has already happened in history, for example, what occurred in former East Germany.[3] However, this social process, along with population losses resulting from diseases, climate changes, and wars, should be welcomed under conditions of scarcity as it will lead to less demand on the social system for all types of resources, as urban communities have always been extremely resource intensive in consumption as I have stated in *World Ecological Degradation*, and Brown (2006) has again recently confirmed. Such tendencies in my view—though they might seem inhuman and barbaric when human lives are at stake—might not be that drastic if human social evolution is to continue. For the good of the many, sacrifices of the few might have to be made. This will occur not through planning and premeditation but as outcomes of the environmental, political, social, and economic crises we discussed in prior chapters.

If localization of economies and polities ensues it will mean that the "deep economy" of McKibben (2007) becomes more of a reality, not by choice but by necessity, for the reproduction of communities. This emphasis on the local means that consumption will rely on easily accessible materials for the reproduction of socioeconomic and political life that are available locally and do not need much energy to access. Local knowledge of the land and conditions comes into play and becomes guides to sustainable living. In this respect, what had been practiced before the arrival of technological advances will increasingly be utilized.

Community assets in terms of skills and knowledge are shared and exchanged. The community replaces the invisible market as the intersection point of exchange processes, thus diminishing the hidden impersonality that now exists between the producer and the consumer. New institutions arise due to diminished resource scarcity. This then is a period of innovation. Alternative forms of exchange can result conditioned by the limits of natural resource availability. McKibben (2007) has highlighted some examples of these potential practices that have been adopted. As a result, the possibility of the resurgence of the community is very likely following the many eras of individualism and atomization of social relations that have characterized our social systems. Such social relations will have an impact on political structures and governance. The centralized political practices will increasingly be replaced by what will work on a local geographic basis depending on the limits mapped out by the landscape and available natural resources. Kunstler (2005) has provided a possible future scenario for the United States.

All in all, it will be a very different world than we have known. Therefore, collapse followed with postcollapse conditions should be welcomed as a period of experimentation, innovation, and transformation. This has happened in the past with the collapse of the late Bronze Age system leading to structural transformations and social experimentation that became part of our social and political organizations even today, albeit with some modifications.

On this "long emergency" to borrow a phrase from Kunstler (2005), which path will be taken is contingent on the severity of the natural, socioeconomic, and political conditions that lie ahead. The catastrophe that has been signaled by many should not be seen as a global calamity. For the privileged it could be. But for the rest, the abyss that we face should be viewed as a period of opportunity. With the slowdown of overall economic activity, population losses, shrinkage of urban areas, and localization processes, this reconfiguration of socioeconomic structures will provide the break for Nature to rest following many years of unsustainable economic expansion. For the human community as a whole, the period opens a window of opportunity for reorganizing and resetting our priorities from what has transpired in the past. Crisis thus provides the occasions for experimentation and innovations. With a lowered population level, a flattened social hierarchy, and deurbanization, there will be less demand on our natural resources. Development instead of growth can be the choice. In this regard, Dark Ages are *Ages of Brightness* for Nature and Us.

If History can be our guide, perhaps there is hope in what is seen as a *hope-less* era. As with past human projects that occurred in world history from the Neolithic through to Urban and Technological Revolutions, social evolution will continue!

NOTES

1. There are other approaches toward mapping out the transformations and possibilities, see, for example, Wallerstein (1998, 2005) and Harvey (2000). For a review of various utopias, see Anderson (2004).

2. The availability of natural resources in these regions has yet to be discovered. According to the United States Geological Survey, one-quarter of the world's undiscovered oil and gas resources are supposed to lie in the Arctic, notwithstanding the marine life under the ice (*New York Times*, October 10, 2005, p. 1).

3. In what is now former East Germany, however, resources from the German government have been plowed back to rebuild and attempt to resuscitate the shrinkage and its social outcomes.

Bibliography

Aberley, Doug. 1998. "Interpreting Bioregionalism." In *Bioregionalism*, edited by Michael McGinnis, 13–42. London: Routledge.
Adam, Barbara. 2004. *Time*. Cambridge, UK: Polity Press.
Adepojo, A. 1995. "Emigration Dynamics in Sub-Saharan Africa." *International Migration* vol. 33 (3/4): 315–90.
Adorno, Theodor et al. 1976. *The Positivist Dispute in German Sociology*. London: Heinemann.
Afolayan, A. A. 2001. "Issues and Challenge of Emigration Dynamics in Developing Countries." *International Migration* vol. 39 (4): 5–36.
Aglietta, Michel. 1979. *A Theory of Capitalist Regulation*. London: Verso.
———. 2007. *La Chine vers la superpuissance*. Paris: Economica.
Ali, Tariq. 2003. *Bush in Babylon*. New York: Verso.
Allen, J. R. M., B. Hientley, and W. A.Watts. 1996. "The Vegetation and Climate of the Northwest Iberia over the Last 14,000 Years." *Journal of Quaternary Science* 11:125–47.
Amacher, G. W. et al. 1998. "Environmental Motivation for Migration: Population Pressure, Poverty, and Deforestation in the Philippines." *Land Economics* vol. 74 (1): 92–101.
Amin, Samir et al. 1982. *Dynamics of Global Crisis*. New York: Monthly Review Press.
Amin, Samir. 1992. *Empire of Chaos*. New York: Monthly Review Press.
Anderson, Perry. 2004. "The River of Time." *New Left Review* (26): 67–77.
Aneesh, A. 2006. *Virtual Migration*. Durham, NC: Duke University Press.
Arrighi, Giovanni. 2005a. "Hegemony Unraveling." *New Left Review* (32): 24–80.
———. 2005b. "Hegemony Unraveling II." *New Left Review* (33): 83–116.
Athanasius. 1980. *The Life of Antony and the Letter to Marcellinus*, translated by R. C. Griegs. New York: Paulist Press.

Bahaguna, Sunderlal. 1996. "Himalaya—From Disaster to Survival." In *Asia—Who Pays for Growth*, edited by Jayant Lele and Wisdom Tettey, 181–90. Aldershot: Dartmouth.

Bao, Y. et al. 2002 "General Characteristics of Temperature Variations in China During the Last Two Millennia." *Geophysical Research Letters* 29 (9): 1324–28.

Barber, Benjamin. 1995. *Jihad vs McWorld*. New York: Ballantine Books.

Baudrillard, Jean. 1975. *For a Critique of the Political Economy of the Sign*. St. Louis: Telos Press.

———. 1983. *Simulations*. Paris: Semiotexte.

———. 1994. *The Illusion of the End*. Stanford: Stanford University Press.

———. 1998. *The Consumer Society*. Beverly Hills: Sage Publications.

Bell-Fialkoff, Andrew, ed. 2000. *The Role of Migration in the History of the Eurasian Steppes. Sedentary Civilization vs Barbarian and Nomad*. London: MacMillan.

Benhabib, Sela. 2002. *The Claims of Culture: Equality and Diversity in the Global Era*. Princeton, NJ: Princeton University Press.

Bentley, Jerry et al., eds. 2005. *Interactions*. Honolulu: University of Hawaii Press.

Berg, Peter. 1978. *Reinhabiting a Separate Country*. San Francisco: Planet Drum Foundation.

——— and Ray Dasmann. 1978. "Reinhabiting California." In *Reinhabiting a Separate Country*, edited by Peter Berg, 217–20. San Francisco: Planet Drum Foundation.

Bergeron, Bryan. 2002. *Dark Ages II When the Data Die*. Upper Saddle River, NJ: Prentice Hall.

Bergmann, Werner. 1985. *Das Frühe Mönchtum als soziale Bewegung. Kölner Zeitschrift für Soziologie und Sozialpsychologie* vol. 37 (1): 30–59.

Bidelux, Robert, and Richard Taylor, eds. 1996. *European Integration and Disintegration: East and West*. London: Routledge.

Black, R. 1998. *Refugees, Environment, and Development*. Harlow: Longman.

Boulding, Kenneth E. 1966. "The Economics of the Coming Spaceship Earth." In *Environmental Quality in a Growing Economy*, edited by Henry Jarrett, 14–25. Baltimore: Johns Hopkins University Press.

Bourke, Austin. 1993. *The Visitation of God*. Dublin: Lilliput Press.

Brandt Commission. 1987. *Common Crisis*. London: Pan.

Bray, R. S. 1996. *Armies of Pestilence: The Effects of Pandemics in History*. Cambridge, UK: Lutterworth.

Bretell, Caroline, and James F. Hollifield. 2000. *Migration Theory: Talking Across the Disciplines*. London: Routledge.

Briffa, K. 1999. "Analysis of Dendrochronological Variability and Associated Natural Climates in Eurasia the Last 10,000 Years." *PAGES Newsletter* 7(1):6–8.

Brown, Lester R. 2006. *Plan B 2.0: Rescuing a Planet Under Stress and a Civilization in Trouble*. New York & London: W. W. Norton & Co.

Brown, N. 2001. *History and Climate Change: A Eurocentric Perspective*. London: Routledge.

Brown, Peter. 1978. *The Making of Late Antiquity*. Cambridge, MA: Harvard University Press.

———. 1989. *The World of Late Antiquity*. New York: Norton.

Burroughs, William J. 2005. *Climate Change in Prehistory*. Cambridge, UK: Cambridge University Press.

Campbell, C. 2004. *The Coming Oil Crisis*. Essex: Multiscience Publishing Co.

Cantor, Norman F. 1994. *The Civilization of the Middle Ages*. New York: Harper.

Carr, Mike. 2004. *Bioregionalism and Civil Society: Democratic Challenges to Corporate Globalism*. Vancouver: University of British Columbia Press.

Carroll, B. A. 1968. *Design for Total War: Arms and Economics in the Third Reich*. The Hague: Brill.

Cartwright, F. E. 1972. *Disease and History*. New York: Crowell.

Castells, Manuel. 1989. *The Informational City*. Oxford, UK: Oxford University Press.

———. 1996. *The Rise of Network Society*. Oxford, UK: Blackwell.

———. 1997. *The Power of Identity*. Boston: Blackwell.

———. 1998. *The End of the Millennium*. Boston: Blackwell.

———. 2000. *The Network Society*, 2nd ed. Boston: Blackwell.

——— and Pekka Himanen. 2002. *The Information Society and the Welfare State: The Finnish Model*. Oxford,UK: Oxford University Press.

——— et al. 2007. *Mobile Communication and Society*. Cambridge, MA: MIT Press.

Castles, Stephen. 2000. "The Impacts of Emigration on Countries of Origin." In *Local Dynamics in an Era of Globalization*, edited by S. Yusuf, W. Wu, and S. Evenett. New York: Oxford University Press.

Castles, Stephen, and Mark J. Miller. 2003. *The Age of Migration*, 2nd ed. New York: Guilford Press.

Chanda, Nayan. 2007. *Bound Together. How Traders, Preachers, Adventurers and Warriors Shaped Globalization*. New Haven, CT: Yale University Press.

Chase-Dunn, C., and E. N. Anderson. 2005. *The Historical Evolution of the World System*. London: Palgrave.

Chernykh, E. N. 1992. *Ancient Metallurgy in the USSR: The Early Metal Age*. Cambridge, UK: Cambridge University Press.

Chew, Sing C. 1992. *Logs for Capital*. Westport, CT: Greenwood Press.

———. 1997. "For Nature: Deep Greening World-Systems Analysis for the 21st Century." *Journal of World-Systems Research* vol. 3 (3): 381–402.

———. 2001. *World Ecological Degradation: Accumulation, Urbanization, and Deforestation 3000BC–AD2000*. Lanham, MD: AltaMira Press/Rowman & Littlefield Publishers.

———. 2002a. "Globalization, Dark Ages, and Ecological Degradation." *Global Society* vol. 16 (4): 333–56.

———. 2002b. "Ecology in Command." In *Structure, Culture, and History: Recent Issues in Social Theory*, edited by Sing C. Chew and David Knottnerus, 217–30. Lanham, MD: Rowman & Littlefield Publishers.

———. 2005a. "From Harappa to Mesopotamia and Egypt to Mycenae: Dark Ages, Hegemonial Shifts, and Environmental/Climate Changes." In *The Historical Evolution of the World Systems*, edited by Christopher Chase-Dunn and E. N. Anderson. London: Palgrave.

———. 2005b. "Globalization, Dark Ages, and Ecological Crisis." In *Changing Face of Globalization*, edited by Samir DasGupta, 48–78. Thousand Oaks, CA, & New Delhi: Sage Publications.

——. 2006a. "Global Ecological Crisis and Ecological Futures." In *Globalization and After*, edited by Samir DasGupta, 184–227. New Delhi: Sage Publications.
——. 2006b. "Dark Ages, Ecological Crisis Phases, and System Transition." In *Globalization and the Environment*, edited by Edward Kick and Andrew Jorgenson, 253–90. New York: Brill.
——. 2006c. "Ecological Crisis Phases and World System Evolution 2200BC–AD1000." In *Globalization and Global History*, edited by Barry Gills and William Thompson, 163–202. London: Routledge.
——. 2007. *The Recurring Dark Ages: Ecological Stress, Climate Changes, and System Transformation*. Lanham, MD: AltaMira Press/Rowman & Littlefield Publishers.
Childe, V. Gordon. 1942. *What Happened in History*. London: Penguin.
Clark, Robert P. 1997. *The Global Imperative*. Boulder, CO: Westview Press.
Climate Science Report. 2006. *Syntheses and Assessment Product 1.1*. Washington, DC: U.S. Climate Science Program.
Constanza, Robert, Lisa Grumlich, and Will Steffen. 2007. *Sustainability or Collapse. An Integrated History and Future of People on Earth*. Boston: MIT Press.
Couch, Chris et al. 2005. "Decline and Sprawl: An Evolving Type of Urban Development—Observed in Liverpool and Leipzig." *European Planning Studies* vol. 13 (1): 117–36.
Cullen, Murphy. 2007. *Are We Rome?* New York: Houghton Mifflin Company.
Daly, Herman, ed. 1973. *Toward a Steady State Economy*. San Francisco: W.H. Freeman.
——. 1996. *Beyond Growth*. Boston: Beacon Press.
Daniel-Rops, H. 2001. *The Church in the Dark Ages*. London: Phoenix Press.
Deffeyes, K. S. 2001. *Hubbert's Peak: The Impending World Oil Shortage*. Princeton, NJ: Princeton University Press.
DeMenocal, P. D. et al., 2000. "Coherent High and Lower-Latitude Climate Variability During the Holocene Warm Period." *Science* (288): 2198–202.
Denemark, R. et al. 2000. *World System History*. London: Routledge.
Dessler, Andrew E., and Edward A. Parson. 2006. *The Science and Politics of Global Climate Change*. Cambridge, UK: Cambridge University Press.
Devall, Bill. 1993. *Living Richly in an Age of Limits*. Salt Lake City, UT: Gibbs Smith.
——. 1998. Humboldt County: Bioregion on the Edge. Unpublished book manuscript, Deep Ecology Resource Center, Trinidad, CA.
Dodds, E. R. 1965. *Pagan and Christian in an Age of Anxiety: Some Aspects of Religious Experience from Marcus Aurelius to Constantine*. Cambridge, UK: Cambridge University Press.
Dodge, Jim. 1981. "Living by Life: Some Bioregional Theory and Practice." *CoEvolution Quarterly* (32): 6–12.
Duckett, Eleanor S. 1990. *Gateway to the Middle Ages: Monasticism*. London: Dorset Press.
Duncan, R. 2001. "World Energy Production, Population Growth and the Road to Oldvai Gorge." *Population and Environment* vol. 22 (5): 503–22.
Dunn, Marilyn. 2000. *The Emergence of Monasticism: From Desert Fathers to the Early Middle Ages*. Oxford, UK: Blackwell.
Eckhardt, William. 1992. *Civilizations, Empires and Wars: A Quantitative History of War*. Jefferson, NC: McFarland.

Ehrlich, Paul, and Anne H. 2004. *One with Nineveh: Politics, Consumption, and the Human Future*. Washington, DC: Island Press.

El-Hinnawi, E. 1985. *Environmental Refugees*. Kenya: UNEP.

Ferguson, Niall. 2004. *Colossus*. New York: Penguin.

———. 2006. *The War of the World*. New York: Penguin.

Frank, Andre Gunder. 1981. *Crisis: In the Third World*. New York: Holmes & Meier.

——— and M. Fuentes. 1989. "Ten Theses on Social Movements." *World Development* vol. 17 (2): 179–92.

Frank, Andre Gunder. 1991. "Transitional Ideological Modes: Feudalism, Capitalism, and Socialism." *Critique of Anthropology* vol. 11 (2): 171–88.

——— and Barry Gills. 1992a. "The Five Thousand Year Old System: An Interdisciplinary Introduction." *Humboldt Journal of Social Relations* vol. 18 (1): 1–55.

———. 1992b. "World System Cycles, Crises, and Hegemonial Shifts." *Review* vol. 15 (4): 621–88.

———. 1993. *World System 500 Years or 5,000 Years*. London: Routledge.

Frank, Andre Gunder. 1993. "Bronze Age World System Cycles." *Current Anthropology* vol. 34 (4): 383–429.

———. 1994. "The World System in Asia: Before European Hegemony." *Historian* vol. 56 (4): 259–76.

———. 1998. *ReOrient: Global Economy in the Asian Age*. Berkeley: University of California Press.

Franklin, C. V., I. Havener, and J. Francis, trans. 1981. *Early Monastic Rules. The Rules of the Fathers and the Regula Orientalis*. Collegeville, MN: Liturgical Press.

Frend, W. H. C. 1972. "The Monks and the Survival of the Eastern Roman Empire in the 5th Century." *Past and Present* (54): 3–24.

———. 1984. *The Rise of Christianity*. Philadelphia: Fortress Press.

Friedman, Jonathan. 2004. "Globalization, Transnationalization and Migration." In *Worlds on the Move*, edited by J. Friedman and Shalima Randeira, 63–90. London: I. B. Tauris.

Gadamer, Hans-Georg. 1975. *Truth and Method*. London: Sheed & Ward.

Gills, Barry K., and William R. Thompson. 2006. *Globalization and Global History*. London: Routledge.

Ge, Q. et al. 2003. "Winter Half-Year Temperature Reconstruction for the Middle and Lower Reaches of the Yellow River and the Yangtze River, China During the Past 2000 Years." *Holocene* 16 (6): 933–40.

Goehring, J. E. 1990. "The World Engaged: The Social and Economic World of Early Egyptian Monasticism." In *Gnosticism and the Early Christian World*, edited by J. E. Goehring, C. H. Hedrick, and J. T. Sanders, 34–42. Sonoma, CA: Polebridge Press.

Gorz, André. 1982. *Farewell to the Working Class: An Essay on Post Industrial Society*. Boston: South End Press.

Green, Henry A. 1986. "The Socioeconomic Background of Christianity in Egypt." In *The Roots of Egyptian Christianity*, edited by Birger A. Pearson and James Goehring, 100–113. Philadelphia: Fortress Press.

Greenspan, Alan. 2007. *The Age of Turbulence: Adventures in a New World*. New York: Penguin.

Griggs, C. W. 1990. *Early Egyptian Christianity from its Origins to 451 C.E.* Leiden: Kluwer.

Grootes, P. M., M. Striver, and J. W. C. White. 1993. "Comparison of Oxygen Isotope Records from Gisp2 and Grip Greenland Ice Cores." *Nature* (366): 552–53.
Habermas, Jürgen. 1974. *Legitimation Crisis*. Boston: Beacon Press.
———. 1979. *Communication and Evolution of Society*. Boston: Beacon Press.
———. 2001. "Why Europe Needs a Constitution." *New Left Review* (11): 5–26.
———. 2006. *Time of Transitions*. London: Polity Press.
Hammer, Thomas. 2004. "Desertification and Migration. A Political Ecology of Environmental Migration in West Africa." In *Environmental Change and Its Implications for Population Migration*, edited by J. D. Unruh, M. S. Krol, and N. Eliot, 231–46. Dordrecht: Kluwer.
Han, Yuhai. 2006. "Assessing China's Reforms." *Economic and Political Weekly* (June 3): 2206–12.
Hansson, Carina Borgstrom. 2003. *Misplaced Concreteness and Concrete Places*. Lund Studies in Human Ecology No. 7. Lund: Lund University.
Hardt, Michael, and Antonio Negri. 2000. *Empire*. Cambridge, MA: Harvard University Press.
———. 2004. *Multitude*. New York: Penguin.
Hardt, Michael. 2002. "Today's Bandung?" *New Left Review* (14) (March–April): 112–118.
Hardy, Edward Rochie. 1952. *Christian Egypt: Church and People*. New York: Oxford University Press.
Harvey, David. 1990. *The Condition of Postmodernity*. Oxford, UK: Blackwell.
———. 2000. *Spaces of Hope*. Berkeley: University of California Press.
———. 2003. *New Imperialism*. Oxford, UK: Oxford University Press.
Hatton, T. J., and Jeffrey Williamson. 2005. *Global Migration and the World Economy*. Cambridge, MA: MIT Press.
Hawken, Paul. 1999. *Natural Capitalism: Creating the Next Industrial Revolution*. Boston: Little Brown & Co.
Heather, Peter, and John Matthews. 1991. *The Goths in the Fourth Century*. Liverpool: Liverpool University Press.
Heather, Peter. 1991. *Goths and Romans*. Oxford, UK: Clarendon Press.
———. 1996. *The Goths*. Oxford, UK: Blackwell.
———. 2006. *The Fall of the Roman Empire: A New History of Rome and the Barbarians*. Oxford, UK: Oxford University Press.
Hill, Christopher, ed. 1996. *The Actors in European Foreign Policy*. London: Routledge.
Hirst, Paul, and Grahame Thompson. 1996. *Globalization in Question*. Oxford, UK: Blackwell.
Hodges, Richard, and David Whitehouse. 1983. *Mohamed, Charlemagne and the Origins of Europe*. Ithaca, NY: Cornell University Press.
Homer-Dixon, T., and J. Blitt. 1998. *Ecoviolence—Links Among Environment, Population and Security*. Lanham, MD: Rowman & Littlefield.
Homeland Security Council. 2006. *National Strategy for Pandemic Influenza Implementation Plan*. Washington, DC.
Hong, Yang et al. 2004. "Environmental-Economic Interaction and Forces of Migration." In *Environmental Change and Its Implications for Population Migration*, edited by J. D. Unruh, M. S. Krol, and N. Eliot, 267–88. Dordrecht: Kluwer.

Hopkins, T., and I. Wallerstein. 1996. *The Age of Transition: Trajectory of the World System*. Atlantic Heights, NJ: Pluto.

Horden, Peregrine, and N. Purcell. 2000. *The Corrupting Sea*. Oxford, UK: Oxford University Press.

House, Freeman. 1998. *Totem Salmon*. Boston: Beacon Press.

Hughes, J. Donald. 2006. *What is Environmental History?* Cambridge, UK: Polity Press.

Huntington, Ellsworth. 1924. *Civilization and Climate*. New haven, CT: Yale University Press.

Huntington, Samuel D. 1996. *The Clash of Civilizations*. New York: Simon & Schuster.

Innes, Matthew. 2000. *State and Society in the Early Middle Ages. The Middle Rhine Valley 400–1000*. Cambridge, UK: Cambridge University Press.

Intergovernmental Panel on Climate Change (IPCC). 2001. *Impacts, Adaptation and Vulnerabilities*. Cambridge, MA: Cambridge University Press.

———. 2007. *Fourth Assessment Report Climate Change 2007*. Geneva: IPCC Secretariat.

International Organization for Migration. 2003. *World Migration 2003 Managing Migration*. Geneva: International Organization for Migration.

———. 2000. *World Migration 2000*. Geneva: International Organization for Migration.

James, Jeffrey. 2003. *Bridging the Global Digital Divide*. Cheltenham: Edward Elgar.

Jameson, Fredric. 2004. "The Politics of Utopia." *New Left Review* (25) (January–February): 135–54.

Johnson, Chalmers. 2000. *Blowback: The Costs and Consequences of American Empire*. New York: Henry Holt.

Jones, A. H. M. 1959. "Overtaxation and Decline of the Roman Empire." *Antiquity* (33): 235–50.

———. 1964 *The Later Roman Empire* AD *244–602*. Oxford, UK: Blackwell.

———. 1974. *The Roman Economy*. Totowa: Rowman & Littlefield.

Kagami, Mitsuhiro et al., eds. 2004. *Information Technology Policy and the Digital Divide*. Cheltenham: Edward Elgar.

Karl, Thomas, Susan Hassol, Christopher Miller, and William Murray. 2006. *Temperature Trends in the Lower Atmosphere*. Report by the U.S. Climate Change Science Program.

Karlen, A. 1995. *Man and Microbes: Diseases and Plagues in History and Modern Times*. New York: Putnam.

Kennedy, Paul. 1988. *The Rise and Fall of Great Powers*. London: Unwin.

Keohane, Robert O., and Stanley Hoffman. 1991a. "Institutional Change in Europe in the 1980s." In *The New European Community: Decision Making and Institutional Change*, edited by R. Keohane and S. Hoffman, 85–105. Boulder, CO: Westview Press.

———. 1991b. *The New European Community: Decision Making and Institutional Change*. Boulder, CO: Westview Press.

Khazanov, Anatoly M. 1984. *Nomads and the Outside World*. Cambridge, UK: Cambridge University Press.

Kinealy, Christine. 1997. *A Death-Dealing Famine. The Great Hunger in Ireland*. London: Pluto Press.

Klein, Naomi. 2001. "Reclaiming the Commons." *New Left Review* (9): 81–89.
Kliot, Nurit. 2004. "Environmentally Induced Population Movements: Their Complex Sources and Consequences." In *Environmental Change and Its Implication for Population Migration*, edited by J. D. Unruh, M. S. Krol, and N. Kliot, 69–99. Dordrecht: Kluwer.
Kristiansen, Kristian. 1993. "The Emergence of the European World System in the Bronze Age: Divergence, Convergence, and Social Evolution during the First and Second Millennia B.C. in Europe." *Sheffield Archaeological Monographs* (6.)
———. 1998a. *Europe before History*. Cambridge, UK: Cambridge University Press.
———. 1998b. "The Construction of a Bronze Age Landscape, Cosmology, Economy, and Social Organization in Thy, Northwestern Jutland." In *Mensch und Umwelt in der Bronzezeit Europas*, edited by Bernhard Hansel, 281–92. Kiel: Oetker-Vosges-Verlag.
———. 2005. "What Language did Neolithic Pots Speak? Colin Renfrew's European Farming-Language-Dispersal Model Challenged." *Antiquity* (79): 679–91.
——— and Thomas Larsson. 2005. *The Rise of Bronze Age Society Travels, Transmission, and Transformation*. Cambridge, UK: Cambridge University Press.
———. 2007. "Household Economy, Long-Term Change, and Social Transformation: The Bronze Age Political economy of Northwestern Europe." *Nature and Culture* vol. 2 (2): 71–86.
Kunstler, James Howard. 2005. *The Long Emergency*. New York: Grove Press.
Lamb, H. 1981. "Climate and Its Impact on Human Affairs." In *Climate and History*, edited by T. M. Wrigley et al., 289–90. Cambridge, UK: Cambridge University Press.
Lawrence, C. H. 1984. *Medieval Monasticism: Forms of Religious Life in Western Europe during the Middle Ages*. London: Longman.
Lipietz, Alain. 1987. *Mirages and Miracles: The Crisis of Global Fordism*. London: Verso.
Liverman, D. M. 1992. "The Regional Impacts of Global Warming in Mexico: Uncertainty, Vulnerability, and Response." In *The Regions and Global Warming*, edited by J. Schmandt and J. Clarkson, 44–68. New York: Oxford.
Lohrmann, R. 2000. "Migrants, Refugees and Insecurity: Current Threats to Peace." *International Migration* vol. 38 (4): 3–22.
Lomborg, Bjørn. 2001. *The Skeptical Environmentalist: Measuring the Real State of the World*. Cambridge, UK: Cambridge University Press.
Lucassen, Jan, and Leo Lucassen, eds. 1997. *Migration, Migration History, History: Old Paradigms and New Perspectives*. Bern: Peter Lang.
Luhmann, Niklas. 1998. *Ecological Communication*. Chicago: University of Chicago Press.
MacDougall, Doug. 2004. *Frozen Earth: The Once and Future Story of Ice Ages*. Berkeley: University of California Press.
Malone, E. E. 1950. *The Monk and the Martyr*. Washington, DC: Catholic University Press.
Mann, Michael. 2003. *Incoherent Empire*. London: Verso.
Manning, Patrick. 2005. *Migration in World History*. New York: Routledge.
Massey, Douglas et al. 1993. "Theories of International Migration: A Review and Appraisal." *Population and Development Review* vol. 19 (2): 431–66.

———. 1994. "An Evaluation of International Migration Theory: The North American Case." *Population and Development Review* vol. 20 (4): 699–751.
———. 1998. *Worlds in Motion: Understanding International Migration at the End of the Millennium.* Oxford, UK: Clarendon Press.
McCormick, M. 2001. *Origins of the European Economy*. Cambridge, UK: Cambridge University Press.
———. 2002. "New Light on the 'Dark Ages': How the Slave Trade Fuelled the Carolingian Economy." *Past and Present* (177): 16–54.
McGinnis, M., ed. 1998. *Bioregionalism.* London: Routledge.
McKibben, Bill. 2007. *Deep Economy. The Wealth of Communities and the Durable Future.* New York: Times Books.
McLeman, Robert, and B. Smit. 2003. *Climate Change, Migration, and Security.* Report 86, Winter. Ottawa: Canadian Security Intelligence Service.
McNeill, William H. 1975. *Plagues and History*. New York: Anchor Books.
——— and Ruth S. Adam, eds. 1978. *Human Migration: Patterns and Policies.* Bloomington: Indiana University Press.
McNeill, William H. 1984. "Human Migration in World Perspective." *Population and Development Review* vol. 10 (1): 1–18.
Meadows, Donella et al. 1972. *The Limits to Growth.* New York: Signet.
———. 2004. *Limits to Growth: The 30-Year Update.* White River Junction, VT: Chelsea Green.
Mesarovic, Mihalo, and Eduard Pestel. 1974. *Mankind at the Turning Point: The Second Report to the Club of Rome.* New York: Dutton.
Meyer, Robert T., trans. 1950. *St Athanasius: The Life of Saint Antony.* New York: Newman Press.
Millennium Ecosystem Assessment. 2003. *Ecosystems and Human Well-Being: A Framework for Assessment*. Washington, DC: Island Press.
———. 2005. *Ecosystems and Human Well-Being*. Washington, DC: World Resources Institute.
Mitchell, William J. 1999. *E-topia.* Cambridge, MA: MIT Press.
Modelski, George, and William Thompson. 1999. "The Evolutionary Pulse of the World System: Hinterland Incursions and Migrations. 4000 BC to AD 1500." In *World System Theory in Practice,* edited by Nick Kardulias, 241–74. Lanham, MD: Rowman & Littlefield.
Modelski, George. 2006. "Ages of Reorganization: Self-Organization in the World System." *Nature and Culture* vol. 1 (2): 205–27.
Mosk, Carl. 2005. *Trade and Migration in the Modern World.* New York: Routledge.
Murphy, Cullen. 2007. *Are We Rome?* New York: Houghton Mifflin.
Murray, W. 1984. *The Change in the European Balance of Power 1938–39.* Princeton, NJ: Princeton University Press.
Myers, Norman. 1993. *Ultimate Security: The Environmental Basis of Political Stability.* New York: W.W. Norton.
——— and Jennifer Kent. 1995. *Environmental Exodus: An Emergent Crisis in the Global Arena.* Washington, DC: Climate Institute.
Myers, N. 1997. "Environmental Refugees." *Population and Environment* vol. 19 (2): 167–82.

Ó Gráda, Cormac. 1999. *Black '47 and Beyond. The Great Irish Famine in History, Economy, and Memories*. Princeton, NJ: Princeton University Press.

O'Lear, S. 1997. "Migration and the Environment. A Review of Recent Literature." *Social Science Quarterly* vol. 78 (2): 606–18.

Olson, Robert, and David Rejeski. 2005. *Environmentalism and Technology of Tomorrow*. Washington, DC: Island Press.

Ong, Aihwa. 2006. *Neoliberalism as Exception: Mutations in Citizenship and Sovereignty*. Durham, NC: Duke University Press.

Organization for Economic Co-operation and Development (OECD). 2000. *Trends in International Migration Annual Report*. Paris: OECD.

Ostwalt, Philipp, and T. Rieniets. 2006. *Atlas of Shrinking Cities*. Ostfildem: Hatze Cantz Verlag.

Parrenas, Rhacel Salazar. 2001. *Servants of Globalization*. Stanford: Stanford University Press.

Parsons, Talcott. 1971. *The System of Modern Societies*. Englewood Cliffs, NJ: Prentice Hall.

Pearson, Birger, and James Goehring, eds. 1986. *The Roots of Egyptian Christianity*. Philadelphia: Fortress Press.

Petras, James. 2002. "Porto Alegre 2002: A Tale of 2 Forums." *Monthly Review* vol. 53 (11): 56–61.

Pirenne, Henri. 1992. *Mohamed and Charlemagne*. New York: Barnes & Noble.

Potts, Lydia. 1990. *The World Labor Market. A History of Migration*. London: Zed Press.

Randsborg, Klav. 1991. *The First Millennium AD in Europe and the Mediterranean*. Cambridge, UK: Cambridge University Press.

Reus-Smit, Christian. 2004. *American Power and World Order*. London: Polity Press.

Rifkin, Jeremy. 2002. *The Hydrogen Economy*. New York: Putnam.

Roberts, Neil. 1998. *The Holocene*. Oxford, UK: Blackwell.

Robinson, William I. 2004. *A Theory of Global Capitalism*. Baltimore: Johns Hopkins University Press.

Rodbell, D. T., G. O. Seltzer, and D. M. Anderson. 1999. "A 15,000 Year Record of El Nino–Driven Alluviation in Southwestern Ecuador." *Science* (283): 516–20.

Rosen, William. 2007. *Justinian's Flea: Plague, Empire, and the Birth of Europe*. New York: Viking.

Rubenson, S. 1995. *The Letters of St. Anthony Monasticism and the Making of a Saint*. Minneapolis: Fortress Press.

Ruddiman, W. F., and J. S. Thomson. 2001. "The Case for Human Causes of Increased Atmospheric Methane over Last 5,000 Years." *Quaternary Science Review* (20): 1769–75.

Ruddiman, William. 2003. "The Anthropogenic Greenhouse Era Began Thousands of Years Ago." *Climate Change* 61 (3): 261–93.

———. 2005. *Ploughs, Plagues, and Petroleum*. Princeton, NJ: Princeton University Press.

Ruggie, John G. 1993. "Territoriality and Beyond: Problematizing Modernity in International Relations." *International Organization* vol. 47 (1): 139–74.

Sachwald, Fredrique. 1993. *European Integration and Competitiveness: Acquisition and Alliances in Industry*. Aldershot: Edward Algar.

Sale, Kirkpatrick. 1991. *Dwellers of the Land*. Philadelphia: New Society Publishers.
———. 1995. *Rebels Against the Future: The Luddites and Their War on Industrial Revolution*. Reading, MA: Addison-Wesley.
Sass, Stephen L. 1998. *The Substance of Civilization*. New York: Arcade Publishing.
Sassen, Saskia. 2006. *Territory, Authority, Rights: From Medieval to Global Assemblages*. Princeton, NJ: Princeton University Press.
Saxenian, Anne Lee. 2006. *The New Argonauts Regional Advantage in Global Economy*. Cambridge, MA: Harvard University Press.
Schäfer, Wolf. 2007. "Knowledge and Nature: History as the Teacher of Life Revisited." *Nature and Culture* vol. 2 (1): 1–9.
Shah, Nasra. 1995. "Emigration Dynamics from and within South Asia." *International Migration* vol. 33 (3/4): 559–625.
Sinor, D. 1977. *Inner Asia and its Contacts with Medieval Europe*. London: Variorum.
Smil, Vaclav. 2004. *China's Past, China's Future*. New York: Routledge.
Snodgrass, Anthony. 1980. *Archaic Greece*. Berkeley: University of California Press.
Snooks, Graeme. 1996. *The Dynamic Society: Exploring the Sources of Global Change*. London: Routledge.
Snyder, Gary. 1974. *Turtle Island*. New York: New Directions Books.
———. 1980. *The Real Work*. New York: New Directions Books.
Speth, James Gustave. 2004. *Red Sky at Morning: America and the Crisis of the Global Environment*. New Haven, CT: Yale University Press.
Soysal, Yasemin. 1994. *The Limits of Citizenship: Migrants and Post National Membership in Europe*. Chicago: University of Chicago Press.
Stein, M. 1998. "The Three Gorges: The Unexamined Toll of Development-Induced Displacement." *Forced Migration* vol. 1 (1): 7–10.
Stern Review. 2007. *Economics of Climate Change*. Cambridge, UK: Cambridge University Press.
Tan, Ming, and Tungsheng Liu et al. 2003. "Cyclic Rapid Warming on Centennial-Scale Revealed by a 2650-Year Stalagmite Record on Warm Season Temperature." *Geophysical Research Letters* 30 (12): 1617–20.
Teggart, F. 1969. *Rome and China*. Berkeley: University of California Press.
Thayer, Robert L. 2003. *Lifeplace: Bioregional Thought and Practice*. Berkeley: University of California Press.
Thomas, R. K. et al. 2006. Temperature Trends in the Lower Atmosphere. U.S. Climate Science Program. Washington, DC.
Thompson, E. A. 1996. *The Huns*. Oxford, UK: Blackwell.
Todd, Emmanuel. 2003. *The Breakdown of American Order*. New York: Columbia University Press.
Toynbee, A. J. 1962. *A Study of History*. Vol. 4. Oxford, UK: Oxford University Press.
United Nations. 2003. *World Population Prospects 2002*. New York: Department of Economics and Social Affairs.
———. 2004. *World Economic and Social Survey 2004 Part 2: International Migration*. New York: United Nations.
United Nations Conference on Environment and Development (UNCED). 1992. *Earth Summit '92*. London: Regency Press.

United Nations Environment Programme. 2002. *World Environment Outlook*. Vol. 3. Nairobi: UNEP.
United Nations High Commissioner for Refugees (UNHCR). 1995. *Statistical Yearbook Populations of Concern to UNHCR: A Statistical Overview*. Geneva: UNHCR.
———. 2006. *The State of the World's Refugees 2006*. Geneva: UNHCR.
———. 2007. *2006 Global Trends: Refugees, Asylum-seekers, Returnees, Internally Displaced and Stateless Persons*. Geneva: UNHCR.
United Nations Population Fund. 2007. *State of the World Population 2007*. New York: UNFPA.
United States Council on Environmental Quality. 1980. *The Global 2000 Reports to the President—Entering the 21st Century*. Vols. 1–2. Washington, DC: Government Printing Office.
Unruh, Jon D., S. Krol Maarten, and Kliot Nurit. 2005. *Environmental Change and Its Implications for Population Migration*. Berlin: Springer-Kluwer.
Varma, Roli. 2006. *Harbingers of Global Change: India's Techno Immigrants in the U.S.* Lanham, MD: Rowman & Littlefield.
Veschuren, D. 2004. "Decadal and Century-Scale Climate Variability in Tropical Africa During the Past 2,000 Years." In *Past Climate Variability Through Europe and Africa*, edited by R. W. Battarbee et al., 139–58. Dordrecht: Springer.
Waever, Ole. 1995. "Identity, Integration and Security: Solving the Sovereignty Puzzle in E.U. Studies." *Journal of International Affairs* vol. 48 (2): 1–43.
Wallace, Claire, and Stola Dariusz. 2001. *Patterns of Migration in Central Europe*. New York: Palgrave.
Wallerstein, Immanuel. 1991. "World System Versus World-Systems: A Critique." *Critique of Anthropology* vol. 11 (2): 189–94.
———. 1998. *Utopistics*. New York: New Press.
———. 2004. *World-Systems Analysis*. Durham, NC: Duke University Press.
———. 2005. "After Developmentalism and Globalization, What?" *Social Forces* vol. 83 (3): 1263–78.
Walters, C. C. 1974. *Monastic Archaeology in Egypt*. Warminster: Aris & Phillips.
Wang, Gung-wu. 1997. "Migration History: Some Patterns Revisited." In *Global History and Migration*, edited by Gung-wu Wang, 1–22. Boulder, CO: Westview Press.
Wang, Hui. 2006. *China's New Order*. Cambridge, MA: Harvard University Press.
White, Lyn Jr. 1967. "The Historical Roots of Our Ecological Crisis." *Science* (155): 1203–7.
Wilson, Edward O. 2002. *The Future of Life*. New York: Alfred A. Knopf.
Wilson, Ernest. 2004. *The Information Revolution and the Developing Countries*. Cambridge, MA: MIT Press.
Wolfram, Herwig. 1988. *History of the Goths*. Berkeley: University of California Press.
Wood, William B. 2001. "Ecomigration: Linkages Between Environmental Change and Migration." In *Global Migrants, Global Refugees*, edited by Aristide Zolberg and Peter M. Benda, 42–61. New York: Basic Books.
Woodbridge, Roy. 2004. *The Next World War. Tribes, Cities, Nations and Ecological Decline*. Toronto: University of Toronto Press.
Workman, H. B. 1913. *The Evolution of the Monastic Ideal*. Boston: Beacon Press.

World Commission on Environment and Development. 1987. *Our Common Future.* Oxford, UK: Oxford University Press.

World Commission on Forests and Sustainable Development. 1999. *Our Forests, Our Future.* Cambridge & New York: Cambridge University Press.

World Economic Forum. 2006. *The Global Information Technology Report 2006–2007.* London: Palgrave-Macmillan.

World Resources Institute. 2005. *State of the World 2005.* Washington, DC: World Resources Institute.

Xu, J. et al. 1996. "On Environmental Migration." *Environmental Science* vol. 17 (3): 81–96.

Zolberg, Aristide, and Peter M. Benda. 2001. *Global Migrants, Global Refugees. Problems and Solutions.* New York: Berghahn Books.

Zook, M. A. 2000. "Internet Metrics: Using Hosts and Domain Counts to Map the Internet Globally." *Telecommunications Policy* vol. 24. (6/7): 232–54.

——. 2001. "Old Hierarchies or New Networks of Centrality?: The Global Geography of Internet Content Market." *American Behavioral Scientist* vol. 44 (10): 1679–96.

Index

About the Author

Sing C. Chew is a professor at Humboldt State University and senior research scientist in the Department of Urban and Environmental Sociology at Helmholtz Centre for Environmental Research—UFZ, Leipzig, Germany. He is the founding editor of the interdisciplinary journal *Nature and Culture*. This book is the final volume of a trilogy on Nature-Culture relations over world history, of which the first two volumes have been published: *World Ecological Degradation* and *The Recurring Dark Ages*.